D1601002

UNCOMMON SENSE

UNCOMMON

N. Metropolis, Gian-Carlo Rota, and
David Sharp, Editors

J. ROBERT OPPENHEIMER

SENSE

Birkhäuser
Boston · Basel · Stuttgart

Copyeditor: Harry Newman

The editors and the publisher would like to thank
Peter Oppenheimer for his cooperation and
assistance in the preparation of this collection.

Library of Congress Cataloging in Publication Data

Oppenheimer, J. Robert, 1904–1967.
  Uncommon sense.

  Includes bibliographical references.
  1. Science—United States—History—Addresses, essays,
lectures. 2. Philosophy—Addresses, essays, lectures.
3. United States—Politics and government—20th century—
Addresses, essays, lectures. I. Metropolis, N. (Nicholas),
1915–   . II. Rota, Gian Carlo, 1932–
III. Sharp, D. H. (David Howland), 1938–   . IV. Title.
Q127.U6066 1984   081   84-439
ISBN 0-8176-3165-8

CIP-Kurztitelaufnahme der Deutschen Bibliothek

Oppenheimer, Julius Robert:
Uncommon sense / J. Robert Oppenheimer, Ed.
by N. Metropolis . . . – Boston ; Basel ;
Stuttgart : Birkhauser, 1984.
  ISBN 3-7643-3165-8 (Basel . . .)
  ISBN 0-8176-3165-8 (Boston . . .)

Contents

Foreword

J. Robert Oppenheimer was born on April 22, 1904 in New York City. Abundantly endowed by nature and circumstance, he was encouraged to pursue any path that captured his interest. Even as a very young man, Oppenheimer delved deeply into literature, the humanities, philosophy and languages, as well as the natural sciences. His sensitivity to moral and aesthetic concerns developed apace with his understanding of the architecture of nature as revealed by science. Thus the preoccupations of the young Oppenheimer foreshadowed one of the most cherished traits of the mature man—his rare ability to interpret, with perception and eloquence, the values and traditions of the humanistic and scientific cultures, one to the other.

In 1925 Oppenheimer graduated from Harvard, summa cum laude. Thereafter he went to the University of Göttingen where he earned his doctorate under Max Born in 1927. In his doctoral research and subsequent research with Born, Oppenheimer made important contributions to the then new quantum theory. Later research was to include contributions of decisive and enduring importance to such diverse branches of modern physics as astrophysics, the quantum theory of molecules, and elementary particle physics.

He returned to the United States in 1929 to accept teaching positions at the California Institute of Technology and the University of California at Berkeley. Oppenheimer was an inspiring and captivating teacher whose concern for students went beyond the purely professional. He worked closely with his students and always seemed to know how to help them reach their greatest potential. For over a decade, these qualities drew many of the best young American theoretical physicists to Oppenheimer, who was thus singularly responsible for guiding to maturity a generation of students who were to lead American physics to unprecedented levels of achievement.

In the spring of 1942 Oppenheimer was asked to join the wartime effort to develop an atomic bomb. During that summer he called together a small group of theoretical physicists at Berkeley and participated with them in an intensive study of the properties of nuclear explosions (considering both fission explosions and the possibility of using thermonuclear reactions) and of the problems which would have to be solved. When it was decided to establish a new laboratory at Los Alamos early in 1943 to design and build the first atomic weapons, Oppenheimer was chosen to be director.

The work on the atomic bomb had been conducted with the utmost secrecy, so that it was only after it had been used that the existence of such a development—with all its tremendous power and significance—became known throughout the country and the world. There was urgent need to inform members of the government and the American people of the nature of the new developments in the field of atomic energy and to assess their effects on national and international policies. Oppenheimer was renowned for the incisiveness and speed with which he analyzed problems and for the precision and elegance of language with which he formulated opinions and conclusions, so that he was immediately overwhelmed with demands by the executive and congressional branches of the government to provide advice on atomic energy matters: domestic legislation, classification policy, the U.S. proposals to the United Nations, and others. When it was first formed, he was appointed by the President to the General Advisory Committee of the U.S. Atomic Energy Commission, and he served as its Chairman from 1946 till 1952. In this role, as well as in response to requests by other agencies of the government, he gave freely of his time and effort to assist with the questions which arose concerning atomic energy.

Oppenheimer's unique service to the government was terminated in 1953 by its decision to suspend his security clearance and to conduct the infamous and tragic Security Board Hearings of 1954.

In 1947, Oppenheimer had been named Director of the Institute for Advanced Study. There, for nearly 20 years, he continued his career as teacher, scholar and valued critic of new developments in theoretical physics. He was responsible for bringing together outstanding scholars from many disciplines, and during his tenure the Institute developed into a leading center for theoretical physics.

For his wartime services, Oppenheimer was awarded the Medal of Merit in 1946. Just ten years after the government's astounding decision concerning Oppenheimer in 1953, President Kennedy selected him to receive the Enrico Fermi Award of 1963. This Award constitutes the highest recognition bestowed by the United States for scientific achievement. It was granted Oppenheimer for his contributions to science and to the community of scientists, and for his dedicated and enormously effective services to the nation. In accepting the Fermi Award from President Johnson, Oppenheimer made no reference to his personal ordeal, but spoke of Jefferson's assessment of the "brotherly spirit of science, which unites into one family all its votaries of whatever grade and however widely dispersed throughout the different quarters of the globe."

Robert Oppenheimer died in Princeton, New Jersey, in 1967.

We have collected here some of Oppenheimer's opinions on science, society and mankind, expressed over the years in his numerous public lectures and addresses. About half of the material reproduced here is previously unpublished. Several years after his death, the themes that concerned him throughout his life are still the concern of our day.

The Editors

Robert Oppenheimer's hat–constant companion and famous symbol.
(OMC)

1

TRAVELLING TO A LAND WE CANNOT SEE

1948

I

One day in a clearing in the forest, Confucius came upon a woman in deep mourning, wracked by sorrow. He learned that her son had just been eated by a tiger; and he attempted to console her, to make clear how unavailing her tears would be, to restore her composure. But when he left, he had barely reentered the forest, when the renewed sounds of weeping recalled him. "That is not all," the woman said. "You see, my husband was eaten here a year ago by this same tiger." Again Confucius attempted to console her and again he left only to hear renewed weeping. "Is that not all?" "Oh, no," she said. "The year before that my father too was eaten by the tiger." Confucius thought for a moment, and then said: "This would not seem to be a very salutary neighborhood. Why don't you leave it?" The woman wrung her hands. "I know," she said, "I know; but, you see, the government is so excellent."

THIS WRY TALE COMES TO MIND often when one observes the efforts which the Government of the United States is making to turn the development of atomic energy to good ends, and the frustrations and sorrows of the negotiations within the United Nations Atomic Energy Commission to which these efforts toward international control have now been reduced.

1

In these notes I should like to write briefly of some of the sources of United States policy, and of the formulation of that policy in the context of the contemporary world. Against the background of present prospects, which manifestly make success in any short term seem rather unlikely, to write of these matters today must of necessity be difficult. We are beyond advocacy, and not yet far enough for history. Yet the effort may not be without some slight usefulness in helping us to achieve an appreciation of what was sound, what was timely, and what was lasting in the policy adopted by the United States, and even more than that, in helping us to see why this policy has not been successful. To answer simply that we have failed because of noncooperation on the part of the Soviet Government is certainly to give a most essential part of a true answer. Yet we must ask ourselves why in a matter so overwhelmingly important to our interest we have not been successful; and we must be prepared to try to understand what lessons this has for our future conduct.

Clearly, such understanding must depend in the first instance on insight into the nature and sources of Soviet policy, and indeed into our own political processes. Such an analysis, which in any final sense may transcend the collective wisdom of our time, is of course wholly beyond the scope of this paper. These notes are concerned solely with questions of our intent with regard to atomic control, questions which, though necessarily overabstract, are yet a part of history.

II

The development of atomic energy had none of the other-worldliness normally characteristic of new developments in science. It was marked from the very first by an extreme self-consciousness on the part of all participants, which has given it an often heroic, though not infrequently rather comic, aspect. Thus when the phenomenon of fission was discovered by Hahn, after less than a decade of intensive exploration of nuclear structure and nuclear transmutations, we were all very quick to hail it, not as a beautiful discovery, but as a likely source of a great technological development. Long before it was known that conditions could be realized for maintaining a fission chain reaction, long before the difficulties in that enterprise were appreciated or methods for their solution sketched out, the phenomenon of fission was greeted as a possible source of atomic explosives, and their development was urged upon many governments. Thus, it happened that when, in the United States, the Manhattan District was approaching the completion of its task, and atomic weapons were in fact almost ready for use, there was a fairly well-informed group of people who in a sort of fraternal privacy had discussed what these developments might mean—what problems they would raise, and along what lines the solution might be sought. After the use of the weapons at the end of the war, much of this thinking became public; it achieved a sort of synoptic codification because of the joint requirements of easy comprehension and military security.

Yet it should be not without usefulness now to recall how the problem appeared to us in the summer of 1945, when it became fully apparent that atomic weapons and the large-scale release of atomic energy were not only realizable, but were about to be realized. Even at that time a good deal of thought had gone into what subsequently came to be known as the peaceful use of atomic energy. On the technical side this preoccupation was natural enough, since many interesting avenues of exploration had been sealed off by the overriding requirements of the military program, and we were naturally curious to sketch out what might lie along these avenues against that time when there should be leisure for their pursuit.

But beyond that there was a political consideration. It was clear to us that the forms and methods by which mankind might in the future hope to protect itself against the dangers of unlimited atomic warfare would be decisively influenced just by the answer to the question "Is there any good in the atom?" From the first, it has been clear that the answer to this question would have a certain subtlety. The answer would be "yes," and emphatically "yes," but it would be a "yes" unconvincing, conditional, and temporizing compared to the categorical affirmative of the atomic bomb itself. In particular, the advantages which could come from the exploitation of atomic energy do not appear to be of such a character that they are likely to contribute in a *very short term* to the economic or technical well-being of mankind. They are among the long-range goods. Thus they could not be expected to

3

recommend themselves as urgent to the peoples of countries devastated by war, suffering from hunger, poverty, homelessness, and the awful confusion of a shattered civilization. The import- ance of these limitations was perhaps not adequately recognized as a deterrent to the sort of interest in the development of atomic energy on the part of other peoples and other governments which might have played so great a part in assuring their support for our hopes. Only among the professional scientists, for whom the interest in the *development* of atomic energy is rather immediate, could we have expected to find, and did we in fact find, an enlightened enthusiasm for cooperation in this development.

Only two classes of peaceful applications of atomic energy were then apparent. To the best of my knowledge, only two are apparent today. One is the development of a new source of power; the other is a family of new instruments of research, investigation, technology and therapy.

Of the former, it was clear two years ago, and it is clear today, that although the generation of useful power from atomic sources would assuredly be a soluble problem and would under favorable circumstances make decisive progress within a decade, the question of the usefulness of this power, the scale on which it could be made available, and the costs and general economic values, would take a long time to answer. As we all know, the answers depend on the raw material situation—essentially, that is, on the availability and cost of natural uranium and thorium—and on the extent to which one could in practice manage to consume the abundant isotope of uranium and thorium as nuclear fuels. Thus, no honest evaluation of the prospects of power in 1945 could fail to recognize the necessity of intensive development and exploration. Equally, no honest evaluation could give assurances as to the ultimate outcome beyond those general assurances which the history of our technology justifies. Certainly no evaluation at that time, nor for that matter today, could justify regarding atomic power as an immediate economic aid to a devastated and fuel-hungry world, nor give its development the urgency which the control of atomic armaments would be sure to have once the nature and ferocity of the weapons had been made clear to all.

With regard to the use of tracer materials, of radioactive species, and of radiations for science, the practical arts, for technology and medicine, we were in a better position to judge what might come.

4

The use of tracer materials was not new. The last decade—the 1930's—had seen increasingly varied and effective applications of them. The use of radiation for the study of the properties of matter, for diagnosis, and for therapy was likewise not new. Several decades of hopeful and bitter experiences gave us some notion of the power and limitations of these tools. What was held in store by the development of atomic reactors and of new methods for the handling of radioactive materials and the separation of isotopes, was a much greater variety and a vastly greater quantity of tracer materials, and a far higher intensity of radiation than had been available in the past. That this would be a stimulus to physical and biological study was clear; that its value would in the first instance depend on the skillful development of chemical, physical, and biological techniques, and that this development even under the circumstances would be a gradual and continuing one, we knew as well.

Thus, our picture of the peaceful uses of atomic energy was neither trivial nor heroic: on the one hand, many years, perhaps many decades, of development—largely engineering development—with the purpose of providing new sources of power; on the other hand, a new arsenal of instruments for the exploration of the physical and biological world, and in time, for their further control, to be added to the always growing arsenal of what scientists and engineers have had available.

Three other matters were clear at that time. On the one hand, the development of atomic power could not be separated from technological development essential for and largely sufficient for the manufacture of atomic weapons. On the other hand, neither the development of power nor the effective and widespread use of the new tools of research and technology could prosper fully without a very considerable openness and candor with regard to the technical realities—an openness and candor difficult to reconcile with the traditional requirements of military security about the development of weapons of war. To these general considerations we should add again: although the peaceful use of atomic energy might well challenge the interest of technical people, and appear as an inspiration to statesmen concerned with the welfare of mankind, it could not make a direct appeal to the weary, hungry, almost desperate peoples of a war-ravaged world. Such an appeal, if made, could hardly be made in honesty.

III

Important though these views as to the peaceful future of atomic energy may have been, they were overshadowed then as they have been overshadowed since by a preoccupation of quite another sort. In an over-simplified statement, this is the pre-occupation for the "control of atomic energy to the extent necessary to prevent its use for destructive purposes." Two sorts of considerations bear on this problem, one deriving from the nature of atomic armament, and the other from the political climate of the postwar world. The former set of arguments has perhaps been given more relative weight in public discussions. Surely it is in the latter that the essential wellsprings of policy should have lain.

Even the weapons tested in New Mexico and used against Hiroshima and Nagasaki served to demonstrate that with the release of atomic energy quite revolutionary changes had occurred in the techniques of warfare. It was quite clear that with nations committed to atomic armament, weapons even more terrifying, and perhaps vastly more terrifying, than those already delivered would be developed; and it was clear even from a casual estimate of costs that nations so committed to atomic armament could accumulate these weapons in truly terrifying numbers.

As the war ended, adequate defenses against the delivery of atomic weapons almost certainly did not exist. There would be variations, as military developments progressed, in the advantages of offense and defense. If effective antiaircraft interception is developed before new types of aircraft or rockets, there may even be periods during which the delivery of atomic weapons is seriously handicapped. But it was clear then that for the most part the development of these weapons had given to strategic bombardment—that form of warfare which peculiarly characterized the last war, and contributed so much to the desolation of Europe and Asia—a new and important and *qualitative* increase in ferocity. It was not necessary to envisage novel and ingenious methods of delivery, such as the suitcase and the tramp steamer, to make this point clear to us. To this must be added a preoccupation not unnatural for us in the United States. It seemed unreasonable to suppose that any future major conflict would leave this country as relatively unharmed as had the last two wars and as totally unscathed by strategic bombardment. These points have been so commonly made, and with such fervor, that they have perhaps obscured to some extent the true nature of the issues involved in the international control of atomic energy.

In this last war, the fabric of civilized life has been worn so thin in Europe that there is the gravest danger that it will not hold. Twice in a generation, the efforts and the moral energies of a large part of mankind have been devoted to the fighting of wars. If the atomic bomb was to have meaning in the contemporary world, it would have to be in showing that not modern man, not navies, not ground forces, but war itself was obsolete. The question of the future of atomic energy thus appeared in one main constructive context: "What can be done with this development to make it an instrument for the preservation of peace and for bringing about those altered relations between the sovereign nations on the basis of which there is some reason to hope that peace can be preserved?"

Although this may have been the question in principle, a far more concrete and immediate problem faced the world. It is true that there may be a certain myopia in too great a preoccupation with relations between the Soviet Union and the United States. It is true that other sources of conflict, other possibilities of war, and other problems which must be solved if the world is to achieve peace can well be discerned and could well be decisive. But although the cooperation—on a scale, with an intimacy and effectiveness heretofore unknown—between the Soviet Union and the United States may not be sufficient for the establishment of peace, it clearly was necessary. Thus, the question naturally presented itself whether the cooperative control and development of atomic energy might not play a unique and decisive part in the program of establishing such cooperation. Clearly, quite widely divergent views might be held as to the readiness of the Soviet Union to embark on such cooperation—varying from the belief that it would be forthcoming if the United States indicated the desire for it to the conviction that it was not in our power to bring it about. The prevalent view, and, I believe, that on which our subsequent policy was based, was that such cooperation would represent a reversal of past Soviet policy, and to some extent a repudiation of elements of Soviet political theory, very much more incisive in fact than the corresponding attitudes on our part. The prevalent view, that is, saw in the problems of atomic energy, not an opportunity to allow the leaders of the Soviet state to carry out a policy of international cooperation, of openness, candor, and renunciation of violence to which they were already committed; rather, it saw an opportunity to cause a decisive change in the whole trend of Soviet policy, without which the prospects of an assured peace were indeed rather gloomy, and which might well be, if accomplished, the turning point in the pattern of international relations.

Why did the field of atomic energy appear hopeful for this endeavor? It appeared hopeful only in part because of the terrifying nature of atomic warfare, which to all peoples and some governments would provide a strong incentive to adapt themselves to a changing technology. As such, atomic weapons were only a sort of consummation of the total character of warfare as waged in this last world war, a sort of final argument, if one were needed, a straw to break the camel's back. But there were other points far more specific. The control of atomic weapons always appeared possible only on the basis of an intensive and working collaboration between peoples of many nationalities, on the creation (at least in this area) of supra-national patterns of communication, of work and of development. The development of atomic energy lay in an area peculiarly suited to such internationalization, and in fact requiring it for the most effective exploitation, almost on technical grounds alone. The development of atomic energy lay in a field international by tradition and untouched by preexisting national patterns of control. Thus the problem as it appeared in the summer of 1945 was to use our understanding of atomic energy, and the developments that we had carried out, with their implied hope and implied threat, to see whether in this area international barriers might not be broken down and patterns of candor and cooperation established which would make the peace of the world.

It was impossible even at that time not to raise two questions of some gravity. One was whether Soviet policy had not already congealed into almost total non-cooperation. The difficulties during the war years, both in cooperation on technical problems which had some analogy to atomic energy, and in the more general matters of the coordination of strategy, could certainly be read as a bad augury for a cooperative future. A second and related question was whether the development of atomic weapons by Great Britain, Canada, and the United States, and the announcement of this completed development at the end of the war, might not itself appear to cast a doubt upon our willingness to cooperate in the future with allies with whom we had not in this field been willing to cooperate during the war.

In any case, these doubts pointed rather strongly to the need for discussions between the heads of state and their immediate advisors, in an attempt to reopen the issue of far-reaching cooperation. The later relegation of problems of atomic energy to discussions within the United Nations, where matters of the highest policy could only be touched upon with difficulty and clumsily, would appear to have prejudiced the chances of any genuine meeting of minds.

In the field of atomic energy our own security demanded a quite new approach to international problems. The security of all peoples would be jeopardized by a failure to establish new systems of openness and cooperation between the nations; and many favorable circumstances made concrete cooperative action appear attractive and feasible. Thus atomic energy had a special role in international affairs. Yet it should be stressed again that no prospect of intimate collaboration in this field appeared likely of success unless coupled with a comparable cooperation in other fields. It should be stressed again that if atomic energy appeared of some importance as an international issue, it was precisely because it was not entirely separable from other issues, precisely because what was done in that field might be prototypical of what could be done in others, and precisely because we appeared to have in this a certain freedom of maneuver—which our technical developments appeared to have given us—to ask for a consideration on the highest possible plane of the means by which the nations of the world could learn so to alter their relations that future wars would no longer be likely.

IV

The views which have just been outlined no doubt reflect only roughly those current in the closing months of the war, among the

9

people to whom familiarity or responsibility had made the nature of atomic energy apparent. That considerations such as these should have found expression in the policy of the people and the Government of the United States is itself somewhat surprising. One must bear in mind that the field of atomic energy was quite unfamiliar to the people of this country, that the whole spirit and temper of a development of this kind would require explaining and reexplaining. One must bear in mind that for reasons of security much that was relevant to an understanding of the problem could not be revealed and cannot be revealed today. One must bear in mind that with the end of the war there was a widespread nostalgia among all our people that the efforts and tensions of the war years be relaxed and that we return to a more familiar and less arduous life. That under these circumstances the United States should have developed, and in large part committed itself to, a policy of genuine internationalization of atomic energy, and that it should have fortified this policy with concrete, if sketchy, proposals as to how the internationalization was to be accomplished, and indeed that it should have taken the initiative in putting these views before the governments of the other Powers—this should not be too lightly dismissed as a remarkable achievement in the democratic formulation of public policy. Nevertheless, this has cost something.

Perhaps most of all what it has cost is that in our preoccupation with determining and clarifying our own policy, we have given far too little thought to attempting to influence that of the Soviet Union on the only plane where such an influence could be effective. We have allowed our own internal preoccupations to make us content to put forward our views in the world forum of the United Nations, without pursuing early enough, on a high enough plane, or with a fixed enough resolution, the objective of making the heads of the Soviet state in part, at least, party to our effort. Our internal effort has cost delay, it has cost confusion, it has cost the injection of some irrelevant and some inconsistent elements in our policy with regard to atomic energy. Above all, it has cost a sort of schizophrenic separation of our dealings in this field from our dealings in all others. In fact, in order to keep pace with political developments all over the world, we have found ourselves negating in many concrete political contexts the possibility of that confidence and that cooperation which we were asking for in the field of atomic energy. It is surely idle to speculate, as it may well be meaningless to ask, whether, if this country had had its own thoughts in better order in June 1945, and had been prepared to act upon them, its policies would have met with greater success. Only an historian to whom the intimacies of Soviet thinking and Soviet decision are freely available would be able to begin to answer such a question. But the evidence, as the actual course of events has unfolded it, necessarily gives little support to the view that by prompter, clearer and more magnanimous action we might have achieved our purposes.

The history of the development of United States atomic energy policy from the first pronouncements of President Truman and Secretary Stimson on August 6, 1945, to the most recent detailed working papers of the United States representative on the United Nations Atomic Energy Commission, is of public record, and has in large part been summarized by the State Department's report "International Control of Atomic Energy."[1] Two aspects of this development need to be specially mentioned. One has to do with what may be called the aim of United States policy—the sketch of our picture of the world as we would like to see it in so far as atomic energy was concerned. Here, the principles of internationalization, openness, candor, and the complete absence of secrecy, and the emphasis on cooperative, constructive development, the absence of international rivalry, the absence of legal right for national governments to intervene—these are the pillars on which our policy was built. It is quite clear that in this field we would like to see patterns established which, if they were more generally extended, would constitute some of the most vital elements of a new international law: patterns not unrelated to the ideals which more generally and eloquently are expressed by the advocates of world government. It has naturally taken some time for it to be clear that more modest attempts at control were likely to aggravate rather than alleviate the international rivalries and suspicions which it is our purpose to abolish.

This solution which the United States has proposed and advocated is a radical solution, and clearly calls for a spirit of

---

[1]"International Control of Atomic Energy: Growth of a Policy." Washington: Department of State Publication 2702, 1947.

11

mutual confidence and trust in order to give it any elements of substance. Only in the field of sanctions—of the enforcement of undertakings with regard to atomic energy—has the policy of the United States necessarily been somewhat conservative. Here in an effort to fit this problem of enforcement into the preexisting structure of the United Nations it has had to rely on the prospects of collective security to protect complying states against the deliberate efforts of another state to evade controls, and to arm atomically.

The second aspect of our policy which needs to be mentioned is that while these proposals were being developed, and their soundness explored and understood, the very bases for international cooperation between the United States and the Soviet Union were being eradicated by a revelation of their deep conflicts of interest, the deep and apparently mutual repugnance of their ways of life, and the apparent conviction on the part of the Soviet Union of the inevitability of conflict—and not in ideas alone, but in force. For these reasons, the United States has coupled its far-reaching proposals for the future of atomic energy with rather guarded reference to the safeguards required, lest in our transition to the happy state of international control we find ourselves at a marked relative disadvantage.

Many factors have contributed to this background of caution. There appears to be little doubt that at the present time our unique possession of the facilities and weapons of atomic energy constitute military advantages which we only reluctantly would lay down. There appears to be little doubt that we yearn for the notion of a trusteeship, more or less as it was formulated by President Truman in his Navy Day address of late 1945: we would desire, that is, a situation in which our pacific intent was recognized and in which the nations of the world would gladly see us the sole possessors of atomic weapons. As a corollary, we are reluctant to see any of the knowledge on which our present mastery of atomic energy rests, revealed to potential enemies. Natural and inevitable as these desires are, they nevertheless stand in bleak contradiction to our central proposals for the renunciation of sovereignty, secrecy and rivalry in the field of atomic energy. Here again, it is no doubt idle to ask how this country would have responded had the Soviet Union approached the problem of atomic energy control in a true spirit of cooperation.

Such a situation presupposes those profound changes in all of Soviet policy, which in their reactions upon us would have altered the nature of our political purposes, and opened new avenues for establishing international control, unfettered by the conditions which in the present state of the world we no doubt shall insist upon. Nor should it be forgotten that were there more reality to the plans for the internationalization of atomic energy, we ourselves, and the governments of other countries as well, would have found many difficulties in reconciling particular national security, custom, and advantage with an overall international plan for insuring the security of the world's peoples. That these problems have not arisen in any serious form reflects the lack of reality of all discussions to date.

Yet despite these limitations the work of the United Nations Atomic Energy Commission has established one point: through many months of discussion, under circumstances of often dispiriting frustration, and by delegates not initially committed to it, the basic idea of security through international cooperative development has proven its extraordinary and profound vitality.

V

The view sketched above of the international aspects of the problems of atomic energy is thus a history of high, if not provably unreasonable hope, and of failure. Questions will naturally arise as to whether limited but nevertheless worthy objectives cannot be achieved in this field. Thus, there is the question of whether agreements to outlaw atomic weapons more like the conventional agreements, supplemented by a more modest apparatus for inspection, may not give us some degree of security. Possibly when the lines of political hostility were not as sharply drawn as they are now between the Soviet Union and the United States, we might have tried to find an affirmative answer to this question. Were we not dealing with a rival whose normal practices, even in matters having nothing to do with atomic energy, involve secrecy and police control which is the very opposite of the openness that we have advocated—and under suitable assurances offered to adopt—we might believe that less radical steps of internationalization could be adequate. The history of past efforts to outlaw weapons, to reduce armaments or

13

to maintain peace by such methods gives little encouragement for hopefulness regarding these approaches.

Nor does it seem reasonable to hope, with the world as it now is, and with our policies in fields other than atomic energy as clearly predicated as they now are on conflict (which is not the same as war) with the Soviet Union, that intermediate solutions involving, perhaps, a formal renunciation of atomic armament, and some concession with regard to access to atomic facilities on the part of international inspectors, will appeal to us as useful. They will hardly do so either in the achievement of present security or the later realization of cooperative relations. Indeed, it has come to be the official position of the Government of the United States that palliative solutions along these lines would almost certainly give rise to intensified suspicions and intensified rivalries, whereas they manifestly would lose for us whatever national advantages—and they cannot *a priori* be dismissed as inconsiderable—our prior development and extensive familiarity with atomic energy now give us.

Clearly we may not lightly dismiss consideration of whether there are other approaches to the problem of the international control of atomic energy which have a better chance of contributing to our security. In fact, recent literature is replete with suggestions along these lines. No one aware of the gravity of the situation can fail to advocate what appears to be a hopeful avenue of approach; and no one has a right to dismiss these proposals without the most careful consideration.

It is my own view that none of these proposals has any elements of hopefulness in the short term. In fact, it appears most doubtful if there are now any courses open to the United States which can give to our people the sort of security they have known in the past. The argument that such a course *must* exist seems to be specious; and in the last analysis most current proposals rest on this argument.

This does not mean that on a lower plane, and with much more limited objectives, problems of policy with regard to atomic energy will not arise, even in the international field. Clearly, arrangements that could be established between the Government of the United States and other governments, for the purpose of profitably exploiting atomic energy or of strengthening our relative position in this field, have some sort of bearing on security, and have an important, if not transparent, bearing on the probabilities for the maintenance of peace. But such arrangements, difficult though they may be to determine, and significant though they may be for our future welfare, cannot pretend, and do not pretend, to offer us real security, nor are they direct steps toward the perfection of those cooperative arrangements to which we rightly look as the best insurance of peace. They belong in the same class, in our present situation, as the proper, imaginative, and wise conduct of our domestic atomic-energy program. They are part of the necessary conditions for the long-range maintenance of peace; but no one would for a moment suppose them to be sufficient.

Thus, if we try to examine what part atomic energy may play in international relations in the near future, we can hardly believe that it alone can reverse the trend to rivalry and conflict which exists in the present-day world. My own view is that only a profound change in the whole orientation of Soviet policy, and a corresponding reorientation of our own, even in matters far from atomic energy, would give substance to the initial high hopes. The aim of those who would work for the establishment of peace, and who would wish to see atomic energy play whatever useful part it can in bringing this to pass, must be to maintain what was sound in the early hopes, and by all means in their power to look to their eventual realization.

It is necessarily denied to us in these days to see at what time, to what immediate ends, in what context, and in what manner of world, we may return again to the great issues touched on by the international control of atomic energy. Yet even in the history of recent failure, we may recognize elements that bear more generally on the health of our civilization. We may discern the essential harmony, in a world where science has extended and deepened our understanding of the common sources of power for evil and power for good, of restraining the one and of fostering the other. This is seed we take with us, travelling to a land we cannot see, to plant in new soil.

At Los Alamos around 1945. (LAL)

2

THE OPEN MIND

1948

A FEW WEEKS AGO the president of a college in the prairie states came to see me. Clearly, when he tried to look into the future, he did not like what he saw: the grim prospects for the maintenance of peace, for the preservation of freedom, for the flourishing and growth of the humane values of our civilization. He seemed to have in mind that it might be well for people, even in his small college, to try to take some part in turning these prospects to a happier end; but what he said came as rather a shock. He said, "I wonder if you can help me. I have a very peculiar problem. You see, out there, most of the students, and the teachers too, come from the farm. They are used to planting seed, and then waiting for it to grow, and then harvesting it. They believe in time and in nature. It is rather hard to get them to take things into their own hands." Perhaps, as much as anything, my theme will have to do with enlisting time and nature in the conduct of our international affairs: in the quest for peace and a freer world. This is not meant mystically, for the nature which we must enlist is that of man; and if there is hope in it, that lies not in man's reason. What elements are there in the conduct of foreign affairs which may be conducive to the exercise of that reason, which may provide a climate for the growth of new experience, new insight and new understanding? How can we recognize such growth, and be sensitive to its hopeful meaning, while there is yet time, through action based on understanding, to direct the outcome?

To such difficult questions one speaks not at all, or very modestly and incompletely. If there are indeed answers to be found, they will be found through many diverse avenues of approach—in the European Recovery Program, in our direct relations with the Soviet states, in the very mechanisms by which our policies are developed and determined. Yet you will not find it inappropriate that we fix attention on one relatively isolated, yet not atypical, area of foreign affairs—on atomic energy. It is an area in which the primary intent of our policy has been totally frustrated. It is an area in which it is commonly recognized that the prospects for success with regard to this primary intent are both dim and remote. It is an area in which it is equally recognized that this failure will force upon us a course of action in some important respects inconsistent with our original purposes. It is an area in which the excellence of our proposals, and a record in which we may and do take pride, have nevertheless not managed quite to quiet the uneasy conscience, nor to close the mind to further trouble.

The history of our policy and our efforts toward international atomic control is public; far more important, it has from the first aroused widespread interest, criticism, and understanding, and has been the subject of debates in the Congress and the press, and among our people. There may even be some notion of how, if we had the last years to live over again, we might alter our course in the light of what we have learned, and some rough agreement as to the limits within which alternative courses of action, if adopted at a time when they were still open to us, could have altered the outcome. The past is in one respect a misleading guide to the future: It is far less perplexing.

Certainly there was little to inspire, and nothing to justify, a troubled conscience in the proposals that our government made to the United Nations, as to the form which the international control of atomic energy should take. These proposals, and some detailed means for implementing them, were explored and criticized, elaborated, and recommended for adoption by fourteen of the seventeen member nations who served on the United Nations Atomic Energy Commission. They were rejected as wholly unacceptable, even as a basis for further discussion, by the three Soviet states, whose contributions to policy and to debate have throughout constituted for us a debasingly low standard of comparison.

This September, the Commission made its third, and what it thought its final, report to the General Assembly, meeting in Paris. It recommended to the Assembly that the general outlines of the proposed form of international control be endorsed, that the inadequacy of the Soviet counterproposals be noted, and that the Commission itself be permitted to discontinue its work pending either a satisfactory prior negotiation between the permanent members of the Security Council and Canada, or the finding by the General Assembly that the general political conditions which had in the past obstructed progress had been so far altered that agreement now appeared possible. The Assembly did in fact accept all the recommendations but one. It asked the Commission to continue meeting. In its instructions to the Commission, however, the Assembly failed to provide affirmative indications of what the Commission was to do, or to express any confidence in the success of its further efforts; in fact, one might dismiss this action as no more than an indication of unwillingness on the part of the Assembly to accept as permanent the obvious past failures of the Commission to fulfill its mandate.

Yet we may recognize that more is involved in this action, that we will come to understand in the measure in which the nature and purposes of our own preoccupation with the problem become clearer. In part at least the Assembly asked that this problem of the atom not be let lapse because it touches in a most intimate, if sometimes symbolic, way the profoundest questions of international affairs; because the Assembly wished to reaffirm that these problems could not be dismissed, that these issues could not be lost, whatever the immediate frustrations and however obscure the prospects. The Assembly was in fact asking that we let time and nature, and human reason and good example as a part of that nature, play some part in fulfilling the age-old aspirations of man for preserving the peace.

In any political action, and surely in one as complex and delicate as the international act and commitment made by the United States with regard to atomic energy, far more is always involved than can or should be isolated in a brief analysis. Despite all hysteria, there is some truth to the view that the steps which we took with regard to atomic energy could be understood in terms of the terror of atomic warfare. We have sought to avert this; we have further sought to avert the probable adverse consequences of atomic armament for our own institutions and our freedom. Yet more basic and more general issues are involved, which, though symbolized and rendered critical by the development of atomic energy, are in their nature not confined to it; they pervade almost all the key problems of foreign policy. If we are to seek a clue to the misgivings with which we tend to look at ourselves, we may, I think, find it just in the manner in which we have dealt, in their wider contexts, with these basic themes.

The first has to do with the role of coercion in human affairs; the second with the role of openness. The atomic bomb, born of a way of life, fostered throughout the centuries, in which the role of coercion was perhaps reduced more completely than in any other human activity, and which owed its whole success and its very existence to the possibility of open discussion and free inquiry, appeared in a strange paradox, at once a secret, and an unparalleled instrument of coercion.

These two mutually interdependent ideals, the minimization of coercion and the minimization of secrecy, are, of course, in the nature of things, not absolute; any attempt to erect them as absolute will induce in us that vertigo which warns us that we are near the limits of intelligible definition. But they are very deep in our ethical as well as in our political traditions, and are recorded in earnest, eloquent simplicity in the words of those who founded this nation. They are in fact inseparable from the idea of the dignity of man to which our country, in its beginnings, was dedicated, and which has proved the monitor of our vigor and of our health. These two ideals are closely related, the one pointing toward persuasion as the key to political action, the other to free discussion and knowledge as the essential instrument of persuasion. They are so deep within us that we seldom find it necessary, and perhaps seldom possible, to talk of them. When they are challenged by tyranny abroad or by malpractice at home, we come back to them as the wardens of our public life—as for many of us they are as well wardens of our lives as men.

In foreign affairs, we are not unfamiliar with either the use or the need of power. Yet we are stubbornly distrustful of it. We seem to know, and seem to come back again and again to this knowledge, that the purposes of this country in the field of foreign policy cannot in any real or enduring way be achieved by coercion.

We have a natural sympathy for extending to foreign affairs what we have come to learn so well in our political life at home: that an indispensable, perhaps in some ways *the* indispensable, element in giving meaning to the dignity of man, and in making possible the taking of decision on the basis of honest conviction, is the openness of men's minds, and the openness of whatever media there are for communion between men, free of restraint, free of repression, and free even of that most pervasive of all restraints, that of status and of hierarchy.

In the days of the founding of this republic, in all of the eighteenth century which was formative for the growth and the explicit formulation of our political ideals, politics and science were of a piece. The hope that this might in some sense again be so, was stirred to new life by the development of atomic energy. In this it has throughout been decisive that openness, openness in the first instance with regard to technical problems and to the actual undertakings underway in various parts of the world, was the one single essential precondition for a measure of security in the atomic age. Here we met in uniquely comprehensible form the alternatives of common understanding, or of the practices of secrecy and of force.

In all this I pretend to be saying nothing new, nothing that has not been known to all thoughtful men since the days of Hiroshima; yet it has seldom come to expression; it has been overlaid with other preoccupations, perhaps equally necessary to the elaboration of an effective international control, but far less decisive in determining whether such a control could exist. It is just because it has not been possible to obtain assent, even in principle, even as an honest statement of intent or purpose, to these basic theses that the deadlock in attempting to establish control has appeared so serious, so refractory, and so enduring.

These words have an intent quite contrary to the creation of a sense of panic or of doom. Yet we need to start with the admission that we see no clear course before us that would persuade the governments of the world to join with us in creating a more and more open world, and thus to establish the foundation on which persuasion might so largely replace coercion in determining

21

human affairs. We ourselves have acknowledged this grim prospect, and responded by adopting some of the very measures that we had hoped might be universally renounced. With misgivings—and there ought to be misgivings—we are rearming, arming atomically, as in other fields. With deep misgivings, we are keeping secret not only those elements of our military plans, but those elements of our technical information and policy, a knowledge of which would render us more subject to enemy coercion and less effective in exercising our own. There are not many men who see an acceptable alternative to this course, although there apparently are some who would regard it as proof of the shallowness and insincerity of our earlier renunciation of these ways. But whether, among our own people or among our friends abroad or even among those who are not our friends, these measures which we are taking appear excessive, or on the whole insufficient, they must have at least one effect. Inevitably they must appear to commit us to a future of secrecy, and to an immanent threat of war. It is true that one may hear arguments that the mere existence of our power, quite apart from its exercise, may turn the world to the ways of openness and of peace. Yet we have today no clear, no formulated, no in some measure credible account of how this may come about. We have chosen to read, and perhaps we have correctly read, our past as a lesson that a policy of weakness has failed us. But we have not read the future as an intelligible lesson that a policy of strength can save us.

When the time is run, and that future become history, it will be clear how little of it we today foresaw or could foresee. How then can we preserve hope and sensitiveness which could enable us to take advantage of all that it has in store? Our problem is not only to face the somber and the grim elements of the future, but to keep them from obscuring it.

Our recent election has seemed to touch this deep sense of the imponderable in the history of the future, this understanding that we must not preclude the cultivation of any unexpected, hopeful turnings. Immediately after the election people seemed stirred, less even by the outcome itself, than by the element of wonder; they would tend to say things like: "Well, after this perhaps we need not be so sure that there will be a war." This sense that the future is richer and more complex than our prediction of it, and that wisdom lies in sensitiveness to what is new and hopeful, is perhaps a sign of some maturity in politics.

The problem of doing justice to the implicit, the imponderable, and the unknown is of course not unique to politics. It is always with us in science, it is with us in the most trivial of personal affairs, and it is one of the great problems of writing and of all forms of art. The means by which it is solved is sometimes called style. It is style which complements affirmation with limitation and with humility; it is style which makes it possible to act effectively, but not absolutely; it is style which, in the domain of foreign policy, enables us to find a harmony between the pursuit of ends essential to us, and the regard for the views, the sensibilities, the aspirations of those to whom the problem may appear in another light; it is style which is the deference that action pays to uncertainty; it is above all style through which power defers to reason.

We need to remember that we are a powerful nation.

We need to remember that when the future that we can now foresee deviates so markedly from all that we hope and all that we value, we can, by our example, and by the mode and the style with which we conduct our affairs, let it be apparent that we have not abandoned those hopes nor forsaken those values; we need to do this even while concrete steps, to which we resort to avert more immediate disaster, seem to negate them.

Our past is rich in example. In that other agony, the Civil War, where the foundations of our government were proved and reaffirmed, it was Lincoln who again and again struck true the balance between power and reason. By 1863, the war and the blockade had deepened the attrition of the South. They had also stopped the supplies of cotton to the English mills. Early that year Lincoln wrote a letter to the working men of Manchester. He wrote:

" . . . It is not always in the power of governments to enlarge or restrict the scope of moral results which follow the policies that they may deem it necessary for the public safety from time to time to adopt.

"I have understood well that the duty of self-preservation rests solely with the American people; but I have at the same time been aware that favor or disfavor of foreign nations might have a

23

material influence in enlarging or prolonging the struggle with disloyal men in which the country is engaged. A fair examination of history has served to authorize a belief that the past actions and influences of the United States were generally regarded as having been beneficial toward mankind. I have, therefore, reckoned upon the forbearance of nations . . . "

Fifteen months later, a year before Lincoln's death, the battle had turned. He could say:

" . . . When the war began, three years ago, neither party, nor any man, expected it would last till now. Each looked for the end in some way, long ere today. Neither did any anticipate that domestic slavery would be much affected by the war. But here we are; the war has not ended, and slavery has been much affected— how much needs not now to be recounted . . .

"But we can see the past, though we may not claim to have directed it; and seeing it, in this case, we feel more hopeful and confident for the future . . . "

In such magnanimity even Grant, at Appomattox a year later, looking beyond the bitter slaughter, looking to nature and to time, could speak to Lee: His troops were to keep their horses; they would need them for the spring plowing.

Each of us, recalling our actions in these last critical years, will be able to find more than one instance where, in the formulation or implementation of policy, we have been worthy of this past. Each of us will mourn the opportunities that may seem to him lost, the doors once open and now closed. Not even in critical times can the sense of style, the open mind, be fostered by issuing directives; nor can they rest wholly on soliciting great actions not yet taken, great words not yet spoken. If they were wholly a matter for one man, all could well rest on his wisdom and his sensitiveness—they neither are, nor can, nor should be. The spirit in which our foreign affairs are conducted will in the large reflect the understanding and the desires of our people; and their concrete, detailed administration will necessarily rest in the hands of countless men and women, officials of the government, who constitute the branches of our foreign service, of our State Department, and of the many agencies which now supplement the State Department, at home and abroad. The style, the perceptiveness, the imagination and the openmindedness with which we need to conduct our affairs can only pervade such a complex of organizations, consisting inevitably of men of varied talent, taste and character, if it is a reflection of a deep and widespread public understanding. It is in our hands to see that the hope of the future is not lost, because we were too sure that we knew the answers, too sure that there was no hope.

In 1947 at the Harvard University Commencement at which Robert Oppenheimer was awarded a honorary degree. Oppenheimer is seated on the very left; third from the left: C. C. Marshall; second from the right: J. B. Conant, then President of Harvard University; on his left General O. N. Bradley. (OMC)

3

SCIENCE IN BEING
Research and the
Liberal University

1949

THE ROLE OF RESEARCH in the liberal university I take as a serious subject, which is in fact closely related to the theme of what the possibilities are, what the future is, what the responsibility is in regard to the varied and rich traditions of the United States. I take that view, though I am not sure that I will be able to make it fully clear—I am not even sure that it is quite clear to me—at least in part because it seems to me that the only way to think of research is as science in being, science in becoming, science as it is: an activity rather than a codified result of an activity. When we look at the world today we see what profound, deep, far-reaching changes science has made. Some of these changes are in the conditions of man's life; many of them are in the alteration of the way in which moral problems come to the individual, come to the community of individuals banded together in government. As a trivial example of this, slavery and poverty did not appear to Greek civilization as the same kind of moral problem that they appear to us. They did not appear as evils because it was not clear that it lay within the power of man to abate them without a sacrifice of everything else.

Science also altered, in ways hard for us to fully appreciate, in the nature of our spiritual life, the values by which we judge things. It has introduced standards for determining whether we are honest men. It has introduced standards for giving meaning to questions and to discovering whether we are in agreement about what we are talking about. It has pointed again and again to

27

human fallibility and to the need to find out when one is wrong as the one necessity for a healthy intellectual life.

It is not my purpose to make a definition of research, and certainly not to go into those marginal areas where it is not quite clear whether research is or is not a reality or a possibility. It is even less my meaning to say what one means by a liberal university. I take it that a liberal university is one that devotes at least a part of its energies—a major part of its energies—to teaching people—not to make them professionals, not to give them vocational guidance, but to give them what is called a liberal education. It would take a brave man to say what that is, but it is certainly at least these things: it is certainly the giving to young men and women of a mode of action, a practice, a training, which binds them in a community; which gives them an activity which in itself is good, irrespective of what professional use they will or will not make of it; which gives them new modes of perception, new modes of evaluation, modes which will stand by them in the future, whether they turn out to be highbrows or lowbrows, whether they turn out to be proconsuls, or scientists, or doctors, or farmers.

The learned way to approach this problem would probably be to give a history of the presence of research in a liberal university as a human institution. But I am, happily, not qualified to do that. An almost equally learned way would be to give the comparative morphology of the universities and to say by what per cent the amount of research had increased and how we compared with the universities of Europe, and so on; but I will make only very incidental use of this mode.

I shall, on the contrary, start by acknowledging what appear to me to be the facts as far as this country is concerned, and what elements of paradox there are in the facts. I would like then to go on in a threefold program: first to say some friendly words about the situation; then to say some hostile words about the situation; and then to try to say some friendly words, which are a little more speculative, and I believe of a somewhat deeper character—by that I mean they may have a better chance of being wrong.

The actual state of affairs has two aspects. In the United States at the moment, research is carried out very largely in the liberal universities. There is a magnificent diversity, however. Research is carried out in all kinds of conditions. It is carried out by the federal government; it is carried out in technical schools; it is carried out in institutes; it is carried out in industries; it is even in cases carried out in the Institute for Advanced Study in Princeton, which is like nothing else. But a great deal of it is in the universities; and it is typical that the part that is in the universities is rather the basic research: that is, research that is aimed primarily at increasing our understanding and our knowledge, without too direct a thought of what use this will be in practice. That this is typically a university function is true in the natural sciences; it is true in the mathematical sciences; and I believe it is even more true in those areas, let us say, of anthropology, psychology, and economics which are becoming subject to research.

At the same time, although the liberal universities support research, there is no well-defined way—there is no clear way—in which research is a part of the liberal education. We will talk more of this, for there is certainly a relation between the two. The relationship is not direct. The young fellow who comes to college to get a liberal education probably does not do any research. If he does, it is probably fake. He comes in contact with people who do research, but in a Jekyll and Hyde relationship. He comes in contact with them while they are not doing research, and he is told, or he reads in the alumni bulletin, that they are great scholars.

This paradox of the double purpose is one of the characteristic improvisations of the country. The primary purpose of the college, historically, was to give young men and women a sound education, a liberal education; certainly it started very largely as a religious education. But it was meant to communicate an appreciation of certain forms of literature, a general knowledge of the way the world was, the ability to read a few languages, the ability to understand the Bible, some history, some mathematics, but not too much. It started in a way very remote from this second purpose, which is to provide the home ground, the fountainhead for the discovery of new truths; yet this is what the great universities of the country have come to be.

In Europe, a similar situation obtains, but it is a very much less acute one, because the number of people who go to the European universities is very much smaller. The universities there are not at all conceived for the education of ordinary people. They are conceived primarily as places of education for people who in one way or another will be specialists. This is changing in Europe; and it will change far more. But in this country it has changed already, and my guess is that the same development that has taken place in the high schools is likely to take place in the colleges.

Often the role of research in a university presents an extreme unbalance. There are universities whose research budgets are enormously greater than their academic budgets. There are universities in which the research budget of a single department, subsidized probably by the government, but sometimes privately, may be greater than the budgets of many, many other departments. There are examples as extreme as this: it is a great liberal university that is the only place in the world, as far as I know, that manufactures, under contract with the United States government, atomic bombs. This is an extreme example of the development of research in the university. I have sometimes asked myself whether we can find any analogy to this situation in the practice of the monastic orders that devote a part of their attention and derive part of their sustenance from the making of their private liqueurs.

The state of affairs we have before us then is this: by and large the liberal universities do sustain research, do sustain about the best that is done; they do this on a very broad front, sustaining characteristically what is basic, where the purpose is the acquisition of new knowledge rather than the application of knowledge. This is a part of the function of a liberal university. Another function may be professional training, training that may be related to research, or may, as in the case of law, or in the case of some other practical disciplines, be rather remote from it. Another function is the maintenance of liberal education. Thus one has a great complexity of functions, as in turn one has great complexity in the direct support of research. This is certainly useful in avoiding abuses, in that no one institution has in this country any monopoly on research. Once again, though research and liberal education are joint functions of the university, they are functions which are not manifestly organically related.

Now the affirmative things that come to my mind to say about this setup—and they certainly are not all-inclusive—are in part at least rather deep. The simplest and least deep is that it is apparently a setup that appeals to the people who do research.

Over and over again a man given the choice of pursuing his work at a university or let us say, in an industry or a government laboratory, will prefer to stay at the university. Once in a while he will find himself so bogged down by academic duties that he will leave and, when the situation gets bad enough, the university reforms and makes the balánce between the claims of instruction and research somewhat more favorable. One reason for this, I think, is that in the relatively creative fields, the fields where imagination is involved, in fields where you can't have any guarantee of success, it is nice to be paid for something different from having good ideas. It is nice to be able to get up and say, "I will teach class today and be a genius tomorrow." There has to be a kind of rhythm to it; there has to be a kind of freedom, which corresponds to the fact that a man, if he is really in trouble, may or may not have that fructifying idea, get that point straight which is bothering him. And this is one of the reasons that the more programmatic institutions, where research is all, are less attractive, in spite of the fact that in a university a professor may be called away from doing what he wants to do, either to teach a class, or more characteristically and more unwelcome, to sit on a committee about teaching a class.

The other point which professors make is that contact with students in a classroom is itself something which they find—and this I think an important clue—harmonious with, and useful to, their own researches. This, I believe, is especially true with the social sciences. I have heard man after man in the field of economics, for instance, say that he wouldn't know how to pursue his work if it were not for the fructification of the classroom.

On the institutional side, there are very serious affirmative things to be said. The professor who doesn't know all the answers, who is trying to find out something that is not quite obvious to him, stands helpless before his problem, part of the time; in a certain sense, he is much closer to the student than the man who just teaches; because the man who merely teaches knows all the answers. The experience of the student is to be puzzled, not to understand, to be confused, and gradually to find some sensible order, to get a new idea, to find out that what he had been thinking was wrong; this is a typical experience for the man engaged in research, and it is a typical experience for the student, and this is one point of harmony. This ability of the man who is

31

worrying about a tough problem to be humbled in his own impotence with respect to the world that he is trying to explore—this is certainly a kind of prototype of how we would like to think of a student approaching a new field, even though he may have there the help of others in finding out new truths. This is one of the reasons why it is a common experience that the most inspiring teachers and the best teachers are also people who devote a good deal of their time to their own researches. One finds that although it is not possible to give a theoretical argument why research and education should occur in the same place, a man himself by uniting these two functions will make it manifest that it is a good idea. This does not always work, but, typically, the fellow who has been worrying about what makes a nucleus hang together, or what is the cause of the dark reaction in photosynthesis, or some other really tough scientific problem, comes to his teaching with a respect for learning, with respect for what other people have done before, and also with respect for ignorance—all of which makes him a far more sympathetic teacher than a fellow who is, by profession, a pedagogue.

Sometimes the universities say that they can get good people only by permitting research and thus seem to justify the maintenance of research establishments as a part of the educational machinery as a sort of bribe. This seems to me not an adequate way to look at it.

Of the affirmative points beyond that, the first is that in this very technical world, in which matters of extreme specialization are often matters of life and death, two things are of deep importance if there is to be a healthy public life. One is that the expert, the fellow who has specialized knowledge, should have some sense of community with people who are not experts, that he should be a man like the rest of them; being in a university, dealing with people who are not committed to a highbrow life, is certainly one way and one of the good ways of achieving this. The other is that in one way or another, every citizen will be called on to have a judgment, and perhaps to have an influence, in matters that have a good deal of technical meat to them. And to have lived in a community where science is in the making, to have associated with people who are helping to make it, is surely one of the ways in which a citizen can be sensitive to what is honest and what is right about a technical opinion and be aware of the kind of thing which science is.

Beyond these things—and I think this is the last of the affirmative remarks I can make—there is another: To live in a university where great researches are in progress and where people are constantly learning the things they didn't know before, learning how wrong they were, learning how much more complex, subtle, and interesting the world is than anyone imagined it to be, makes the habit of open-mindedness a natural thing. The habit of not knowing all the answers, the habit of inquiry, the habit—I think research is defined in the dictionary as persistent inquiry—anyway, the habit of occasional inquiry, the notion of intellectual adventure, the discipline of having it manifest that there is such a thing as right and wrong are things which flourish in a community where research is going on. They are a great part of the intellectual and, I believe, the spiritual tradition of our time. There can be little doubt that one of the virtues of the present system is that, in a loose way, young people, who spend four years in college in the hope of coming out of it wiser, better informed, knowing more, and more skilled, live in contact with this part of our intellectual life.

The adverse things that can be said are probably even more familiar. They all stem from the fact that the relation between research and general education—and here we leave out the technical schools and professional education because they are of a lesser order of difficulty—is of a subtle, unmanifest, and dis-organized form. They all come back to the question: what is it to a young man who is taking a series of courses, trying to find out what Plato thought, trying to find out what the laws of economics are alleged to be, trying to find out what it was that made people think that Chaucer was an interesting writer, what is it to him that somewhere in a building off by a canyon, or upon a hill—anyway, with the best view in town—research is going on which costs hundreds of millions of dollars and with which he has really no possibility of coming into contact? He knows it is there.

This brings with it a whole series of evils which I am not going to be able exhaustively to discuss. I think the first of them, the most manifest of them is probably seen in the President's office, where the gross administrative unbalance, caused by trying to ride

33

simultaneously the elephant of research and the mouse of general education, makes a very odd situation. This shows up, for instance, in the often-raised question: aren't we devoting too much emphasis in the selection of our faculty on research ability? Shouldn't we get good teachers? Of course, if there is any health in the situation, you shouldn't have to ask this question, because the two things, by and large, should be coextensive. There should be exceptions, but they should be exceptions. There is the difficulty of status. How are you going to balance the activities in a field where research either is impossible or should be discouraged against the activities in a field where research has enormous prestige, and where an enormous amount of money and high salaries, and so on, can be commanded? How are you going to keep the purpose of the general education big enough in the university's program and honest enough in the university's program so that it will not be submerged by the enthusiasts who collect a few hundred thousand dollars from the Office of Naval Research and a few hundred thousand dollars from another foundation and go off and spend their time away from the students, thereby enriching the life of the university?

This is only one aspect of the trouble; I think that there are far greater dangers. One thing I have myself seen is the many examples of people who, although they like the university climate, abuse it, lose their interest in instruction, in teaching, become specialists, become impatient of the interruption of classes, and go off by themselves and pursue their own studies. A few people of that kind can make a very great difference in a university. To have most people in that status would certainly be to make relations between research and education not only tenuous, but non-existent.

I have listed three further examples of the kind of trouble which research makes in the general education. One of them is in the field of the humanities, in which we may, for instance, think of literature, literature in all languages that a man cares to read, as a typical example. It is not true that there is no research in the humanities that makes sense. There are textual matters which need to be explored. There are matters of tracing influences that are worthy undertakings. But these are of a rather minor kind; and above all, they are, by and large, irrelevant to the reason why the humanities are useful to general education. The reason why the humanities are useful in general education is that, by learning to read, especially by learning to read in a variety of languages and in a variety of times, the student is given a power of action and evaluation which will stand with him; to know that the text that he has read has or has not been challenged is a secondary point, and very largely, an irrelevant one. Yet the pressure of competition and status with the physical sciences, the natural sciences, and ultimately the social sciences, has forced upon many sincere scholars in the humanities a quite irrelevant and quite inappropriate enthusiasm for research. And in this case, I believe, this not only does not help the humanities as an ingredient in general education, but, I believe, it has distracted from their real value. Once again: This is not to say that there is no such thing as research. But it is certainly a small part of what the humanities are good for; and it occupies in literature a role absolutely not of the same order as the role that research occupies in, let us say, astronomy or in mathematics.

One finds in social studies a different and very much more complicated and subtle case. That is a case where research certainly is not impossible, and where the methods of science, in one way or another, are sure to be applied; but where their application has caused a kind of disintegration of disciplines once regarded as unified, which has made it very much harder to make these disciplines useful in the general education of the student. There was a time when economics, history, and political science were not recognized as separate disciplines, when they were all part of what might have been called political economy. I have heard it argued, and although I am very much out of my depth, I think it can be argued, that political economy is a far more adequate theme to teach in general education than are its component parts, in particular, than are political science or economics. The social scientists themselves have recognized the price that is being paid in the disintegration, more or less inevitably consequent on research; research with its abstract and monkeying techniques always breaks things down, always looks at pieces, always isolates and rarifies. And I believe that the attempt to put social science together again, like Humpty Dumpty, characteristically but not solely in what are called area studies, is a symptom of the social scientists' own feeling that something has been lost which is of practical value. My purpose here is not to

35

make a critique of the programs of the social sciences, a task for which I would feel inadequate, and which would certainly take us far beyond the scope and theme of this essay, but only to point out that it is another case where the interests of research are, at first sight at least, not particularly harmonious with the interests of general or liberal education.

A still more striking case you will find in philosophy, which a few centuries ago comprehended most of the subjects which are now the object of research, and which today retreats to a narrower and narrower domain. Perhaps the last really great philosopher of a university, whom I remember, was Whitehead. Yet Whitehead himself started one of the great disintegrations of philosophy by helping to make, and to make popular, the advance of mathematical logic, which now lives in the department of philosophy but is soon going to move out, leaving even less behind, and making an even greater fragmentation. The man who specializes in mathematical logic is not necessarily as likely to be as useful in a liberal education as the general philosopher. He is less likely to have an interest that corresponds to the interest of his students, or to be able to give an insight into what philosophical thought has been in the past and the role it has played, than is the man who has become somewhat less specialized.

As you know, many efforts to put these things back together, to make a dish for the liberal education out of the components which the analytical, experimental, and interfering techniques of science have created, are now under way in universities of the country. I believe they all, generally speaking, are called general education. They all have in common the attempt to apply the historical method, rather than the experimental method of scientific research, to research itself, namely to the progress of learning as a historical process, which may be communicable to the student, even though the objects learned, the things learned, are not communicable to the student. I have grave doubts as to whether this way of interpreting science as a humanity is going to be successful, whether this way will put back together what man has taken apart.

I have tried to mention some of the difficulties of the present situation. All of them could be met by a kind of balance, by saying, "Let us have some research, but let us not have too much; let us have some liberal education, but let us not permit it to interfere with research; let us not permit research to interfere with liberal education, let us have a kind of harmony. Let us not restrict research to the universities, but let us also admit it there." This answer of balance is very likely, in my opinion, to be the one that will be followed. It is the one that I suppose most of us, if forced to the wall, would advocate, to deal with a paradox of multivalent functions in a single human institution. But I would like at least to try out a different view of what the future may hold, not because I am confident that it will, but because I myself wish that it would.

The deeper view is this: research itself, the life of science, is a mode of action. It involves its own values. It involves its own community. And the question we really have to ask is whether this aspect of science is deeply enough rooted in our community, in the political, intellectual, spiritual life of our times so that research will not merely be symbiotic with general education but will be a large part of its substance.

If we look at the ingredients of the liberal education of the past, it was because the student could learn to do something from his teacher that this was included in the curriculum. The saint gave to the priest something which the priest could give to everyone. The poet gave to the man of letters something which the man of letters could give to everyone. The philosopher-king, or, more modestly, the statesman, gave to the professor of moral philosophy something which he could give to all students, as a mode of action, as a way of living, with implicit values. What are the modes of action, what are the characteristics of the values, what is the community, which science in being brings to us?

It is above all a world in which inquiry is sacred and freedom of inquiry is sacred. It is a world in which doubt is not only a permissible thing, but in which doubt is the indispensable method of arriving at truth. It is a world in which the notion of novelty, of hitherto unexpected experience is always with us, and in which it is met by an open-mindedness that comes from having known, of having seen over and over again, that one had a great deal to learn. It is a way of life in which the discovery of error is refined, in which almost all the ingenuity that goes into experimental, analytical, or mathematical techniques is devoted to refining, sharpening, making more effective the way of finding out that you are wrong; this is the element that creates discipline. The

37

nature of the discipline of science is its devotion, its dedication to finding out when you are wrong, to the detection of error.

These are ways that at one time appeared to be of very great relevance to the political life of democracy. You have only to look at the eighteenth century, you have only to read Condorcet, you have only to read Jefferson in our own country, in what he said and wrote of our own constitutional system, to recognize that these values, that these disciplines, characteristic of science, and coupled with the forward-looking character which science always has, were regarded as basic to the functioning of political democracy. You have only to look back there to see how the dread of totalitarianism, the dread of authority, the dread of dogma, was a two-valued thing: on the one hand, the indispensable thing for science, and therefore for progress; and on the other hand, the indispensable condition for democracy, and therefore for freedom.

The absence of dogma, the absence of authority, the fact that even those things in which you most believe are open to doubt—and your willingness to doubt even those things to which you are most committed—are, it seems to me, part of the freedom and responsibility of America, and the challenge to American institutions.

It seems to me that this is a time when the relevance of the virtues of science to ordinary human life, to political life, and even to personal life could hardly be more manifest. Yet I do not know whether the experience of participating in science in being, in one way or another, whether this will be, whether it can be made to be, a sufficiently general experience, whether it can become an inspiring and steadying and unifying feature of our time. The decision, I think, on the role of research in a liberal education depends on just that question. At the least the two will be together in a kind of harmonious symbiosis, of the kind that we have seen in the past years. But at the most, science in being, research, may be to the liberal education, not an accident, not an ancillary or secondary or convenient thing to be held in balance—it may be the scripture itself. We will have then to do, not with a talented, fortunate, social improvis[at]ion, but with a world which hangs together far more than the world in which we have been brought up.

38

These last views are not a prediction, and of their wisdom I myself feel grave doubts. Yet it is in the hope of stimulating not only skepticism, but hope itself, that I have brought them forward.

Robert Oppenheimer and J. Nehru; Kitty Oppenheimer in the background; around 1950. (OMC)

4

THE CONSEQUENCES
OF ACTION

1951

μή, φίλα φνχά, βίον ἀθάνατον σπεῦδε, τὰν δ ἔμπρακτον ἄντλει μαχανάν[1]

[. . .] IN THE RECENT PAST much has changed. Our troops are at war in Korea. We are in a state of emergency, and are mobilizing. Many of the views of the American people have sharpened and altered. Errors that were prevalent six months ago are obvious as errors today. There is a deep anxiety about war, about the prevention and limitation of war, and about the defeat of our enemies should war break out. I thus thought it only right that I should address myself largely, though not exclusively, to the role of the atom in military matters, to the public aspects of this question, of which obviously not all aspects can be or are public. [. . .] This is a field in which there are many handouts and many classified lies, in which the wholesome give and take of question and answer are much needed. Where I can, I shall try to respond to these questions of concern and curiosity.

From the beginning there has been a problem of assimilating the atom into the life of the country, and of making its development useful for our purposes. One may remember the week in August in which we all learned about the development of atomic energy; it was quite a remarkable week. It was generally thought, in that August of 1945, that the war, though sure of a victory, might continue through many months, and with many casualties. In that week, Hiroshima was bombed, the Soviet Union declared war on Japan, Nagasaki was bombed, and the Japanese

---

[1]Pindar, *Pythian III*: "Dear soul, do not pursue immortal life; exhaust the practicable technical resources."

Government made it known that it was prepared to sign an instrument of surrender. In that week, the most terrible war mankind had lived through came to an end; 100,000 people were killed by two atomic bombs, and about as many others were injured.

It was not unnatural that one should try to make of this spectacular development some useful, constructive application to our national life. In the often bizarre efforts to do this, I have been reminded of an old, old story, which surely you all have heard, about the man who stuttered and for therapeutic purposes, was taught to say "Peter Piper," and so on, and complained afterward that although he could say this without stuttering, he could not work it into an ordinary conversation.

We have been engaged in the last five years, in trying to work the atom into an ordinary conversation. There are many odd examples. No one needs today to explain that the atom does not mean world government, nor free power, nor the reform of our educational institutions. But there are some areas where the struggle to reconcile this development with our traditions, our needs, and our intentions has had, it seems to me, some importance or some interest. I want to consider three of those instances, the first two perhaps rather less than the third.

The first is the early post-war attempt to build around the atom new elements of our relations with the Soviet Union, of international relations in general, an attempt which failed almost before it started. The second is the effort [. . .] toward reconciling the administration of the atomic energy program with the traditional processes for maintaining responsibility of our government to our society. The third is the nature of the contribution which atomic energy may reasonably be expected to make to our military power, to our power to prevent war and win it.

In all of these, there is the recurring theme of the reconciliation of novelty and tradition, of things that are new with things that are known. That this would be so was anticipated very early. In October of 1945, my friend, the distinguished economist, Professor Jacob Viner, spoke before the National Academy of Sciences and the American Philosophical Society at a symposium on atomic energy. He quoted a phrase from the President's message to Congress of October 3rd of that year. The phrase reads: "In international relations as in domestic affairs, the release of atomic energy constitutes a new force too revolutionary to consider in the framework of old ideas." Professor Viner remarked: "Beyond a few facts and a few surmises about the military effectiveness and the cost of atomic bombs, however, I unfortunately have no materials to work with except a framework of old ideas, some of them centuries old . . . I suspect that practically every non-scientist is in substantially the same predicament, except that many are unfamiliar even with the old ideas . . . "

Professor Viner need not have exempted scientists from his statement. The problem of dealing wisely with the atom has been precisely this problem of using ideas that we had, and not ideas that we might hope to have, in order to deal with a quite strange new subject.

I

We may think back to 1945. At that time the people and government of this country were concerned with the building of a decent and secure peace. We still are; but the preoccupations which were then natural are, alas, not contemporary today. The efforts which were made to build around the atom the beginnings of a system to secure peace seem to me worth recalling; and I think we will be better off for remembering them, even if they are not ideas whose application today looks immediately hopeful.

Our proposals for the International Control of Atomic Energy, which were largely based on the technical realities of the field, were presented on our behalf to the United Nations by Mr. Baruch, and were widely accepted by the non-Communist nations. The implementation of these proposals would have required a profound alteration in some, at least, of those features of the Soviet system which are responsible for the great troubles we are in today. The failure to persuade the Soviet Government to alter its practices was anticipated by many. Yet we should not forget that this is an objective not only of the past but of the future as well.

[. . .] It was clear that no secure system could be developed for protecting people against the abuse of atomic weapons, unless the world were open to access, unless it was possible to find out the relevant facts everywhere in the world which had to do with the

43

security of the rest of the world. This notion of openness, of an open world, is, of course, relevant to other aspects of the Soviet system. It is doubtful whether, without the newly terrible, yet archaic, apparatus of the Iron Curtain, a government like the Soviet Government could exist. It is doubtful whether the abuses of that government could persist.

Nowhere has there been a more eloquent and more general account of this ideal of an open world, an ideal in which secrecy would not be used for national purposes, in which everything of relevance to the common security and common welfare would be accessible, than in the efforts and the writings of the beloved and eminent physicist, Niels Bohr. If we ever hope to see the world put peacefully together again, it will have to involve, as one of its essential ingredients, an openness with regard to those parts of life which, if held secret, can be a menace to all mankind.

There is another theme which appeared in the United States proposals, which has recurred, and will again. This is the notion of cooperation with other nations in the application of science and technology for the betterment of the conditions of life. It is essentially the theme that has reappeared in the Point IV program. It is an expression, appropriate to our time and our country, of a universal sense of fraternity.

I would like to make two comments: The first is that, if we are to return to these themes, and I hope we will live to do so, they will have to be on a broader basis than in our initial proposals to the U.N. A mistake, which was in no way decisive from the point of view of Soviet objection to them, was that the atom was treated as too special. This is a mistake that we meet again and again when we study the brief history of atomic energy.

Let us consider again the suggestion of a Development Authority for cultivating, for the benefit of all, the affirmative peaceful advantages of atomic energy. That was a good idea; but it was only part of a very good idea. It might have been a large part, had practical atomic power become rapidly available. That has not happened. I think we all know that the Atomic Energy Commission is not producing any power; that on the contrary it is setting about to become the greatest user of power in the world. This very fact will, I hope, provide an incentive to do something at least to reduce the power deficit that the atomic enterprise has become. But it is clear now that only when taken together with other branches of technology, only when the affirmative things derived from the atom are taken as a part of science and technology as a whole, is there anything substantial and robust to develop, anything whose benefits are worth extending to the rest of the world.

There is an analogous need for broadening the control functions of an International Authority. No one today would regard the negative and prohibitive aspects of an agreement on atomic energy as entirely separate from those applying to other arms. When President Truman addressed the General Assembly last autumn, he called attention to both points. He asked that the two commissions on atomic energy and disarmament be combined; and asked that the Point IV program be accepted and expanded. We know to what an unruly and to what a preoccupied world these suggestions were made. I believe that if with an open mind we remember what our objectives were, we will find opportunities to promote them. We will forget them at our peril.

II

With the collapse of the efforts to reach international agreement, it was clear that the atomic energy enterprise in this country would be very large, that it would be largely of military interest, that it would be largely secret, and that it would be largely monopolistic. The question at once arose, in drafting the law, and in administration of the law, whether there could be adequate safeguards, so that in decisions, administrative decisions as well as policy decisions, the powers given to the Commission, and the strange, new, and rather perverse definition of secrecy within which it operated, would not be abused or misused; so that there would be an accountable and responsible administration even under the veil of secrecy, so that the decisions taken would reflect a full awareness and appreciation of all relevant facts.

We have often learned of decisions taken in this field in which all that has appeared in public has been a sort of superficial ripple, and it has not been easily possible to conclude as to whether the decisions were wisely or foolishly taken. Let me give two examples: About a year ago the President said that he was directing the Atomic Energy Commission to proceed with the work on all forms of atomic weapons, including the so-called hydrogen or thermonuclear weapon. Neither the procedures, nor the arguments, nor the consequences of this decision are in the public domain. The other is a decision which apparently was made, perhaps by default, a little over a year and a half ago. Senator McMahon[2] raised the question of how the Congress, how anyone, could have any valid notion of whether the Commission was or was not doing its job, unless they had some idea of what it was producing in the way of atomic weapons. Clearly, a decision was reached not to make this information available; but the reasons and the arguments are again not, to my knowledge, public.

As a digression on the epiphenomena of secrecy, let me cite a public record that has been rather poorly used for public understanding. For the last two years, I have seen many estimates of how many bombs we have, all allegedly deriving from testimony I gave about five years ago before the Special Senate Committee on Atomic Energy, estimates differing widely with the differing arithmetical practices of the reporter. Even in this last week I have heard and seen three such estimates.

Let me cite the relevant excerpts of the testimony.[3] Senator Tydings asked: "Assuming that ten years from now atomic energy in many countries has been licensed by the Government for peacetime manufacture and uses . . . It if were decided to make military bombs from our peacetime atomic energy, how long would it take us to complete 200?" I said, "Maybe a little over a year." Senator Tydings said, "How long would it take us to make 50?" And I said, "Maybe a year." And then I said, "I think a year is too long; maybe nine months."

It is clear that these estimates concern the rate of conversion of fissionable material into weapons, whereas the pacing factor in the making of atomic weapons has for us been the making of fissionable material. Nevertheless, this testimony has been quoted and requoted as an estimate of our weapons stockpiles, perhaps unchallenged—certainly not adequately challenged—for many years.

---

[2] Bulletin of the Atomic Scientists, March, 1949, page 66.
[3] Senate Resolution 179, Vol. 2, page 215.

The Committee of the Bar Association clearly could not address itself to a study of all the errors and confusions deriving from secrecy, nor to a study of decisions in which real elements of secrecy were involved, in which no adequate public record existed, and in which it could, at most, have listened to the unclassified gossip of those who participated. The Committee has instead studied decisions, explanations, and administrative proceedings of a less inflammatory, less spectacular nature. It has tried to track them down in areas where there is a full public record, where there is a detailed account given by the Commission, before other agencies, and in hearings before committees of the Congress.

In this way, the Committee of the Bar Association is trying to answer the question: To what extent—and to what extent inevitably—has there been an abuse of secrecy? To what extent has there been an irresponsible use of the powers given to the Atomic Energy Commission? To what extent is the system working within the framework of responsible, accountable, traditional procedure? I shall not try to report on this work. It is not finished; and one will have a qualified report from the Committee at a later time. This work is another example of the contributions which the law is making to the assimilation of a new field into a tradition that we need to preserve.

That this work will be of special relevance in the months and years ahead is obvious. The work of the Atomic Energy Commission is expanding. It is expanding into a general mobilization. Materials and power are going to be controlled, and one can foresee some major and rather spectacular collisions in the impact of the work of the Commission on the general rearmament program. How these are resolved, whether they are resolved to make the most of the national economy, whether they are resolved to increase the national strength as much as possible, depends almost wholly on whether the management of largely secret and very powerful agencies is responsible.

III

There is another group of problems, in which the assimilation of a new and special field and a tradition has, it seems to me, great,

47

immediate importance. That lies in the contribution which one may reasonably hope that the atom can make to our military power, the power for the prevention of war, the limitation of war, and for the defeat of the enemy in the event that war does come. It is clear that not all the aspects of this problem are public or can be public. What is important is that there are some aspects that are public. I need to make some comment on these.

In the past the debate about the military value of the atom has had a singularly empty quality. To the first impression that the atomic weapon was so great a thing that it was a decisive, an absolute military power, there was a reaction: it is another weapon, it is "just another weapon," or, as in Mr. Hoover's phrase, it is "a less dominant weapon," than we had thought. People close to the work have at times also thought that the atom was a bit of a gold brick. But, in fact, one cannot talk in these terms. This is the argument of the optimist who thinks that this is the best of all possible worlds, and of the pessimist, who knows it. This is not an argument that has meaning.

For our purposes, at this time, there is a very definite thing we need to say: the difficulty and the magnitude of the military and the political problems which we now face and will continue to face, and the extent of our investment in the atomic field, mean that we cannot afford to misuse, and we cannot afford to ignore, what the atom can do for military purposes. This is a luxury in which we should not indulge.

In what I have said, and shall say, I am limiting myself, and I think rightly limiting myself, to one use only of atomic energy, one class of uses, the atomic bomb. There have been many references in the papers to other projects: to poisons, to other kinds of explosives, to propulsion systems for military craft of one kind and another. That is enough to indicate that some technical work has been done on them. But it is not of them I am speaking, but of the atomic bomb.

There are two sides to our problem, though they are related. One side is the technical and the military: questions of what we do to make weapons available, what weapons we make available, how we plan to use them; the other is the side of policy, the conditions under which we might use atomic weapons, their significance in the conduct of war, their significance in international relations. It is clear that these two sides are related; and yet it will be useful, I think, to separate them; for the role of the public is quite different in them.

With regard to the first group of questions, the technical ones, technical both for engineers and scientists and for the military people, the public role is probably in the first instance to determine that secrecy and power are not being abused, that the right questions are being asked and that reasonably honest men are trying to answer them. There is a lot of hard work to do, much of which has not been completely done at the moment. There is, as I have mentioned, and as the Commission has made clear, an increase in the scale on which the explosives are to be manufactured. That will not be a trivial undertaking; for there is not only a problem of a balance of the various ways of making explosives, but of a balance between them and other military efforts, a necessarily tight balance in a period of mobilization. The use of electric power and the use of other scarce materials are examples.

There is an obvious need for the development of weapons systems, so that one can use atomic bombs in a variety of ways, so that one can deliver them in more than one way, and so that one can make them for a variety of targets and uses and situations. There is need for operational planning, so that one may be prepared to anticipate under what conditions they are good weapons, and a good use of explosive, and under what conditions they are not; and there is need for serious work on such countermeasures as exist. Everybody knows that there are no special countermeasures against atomic weapons; but if we can intercept carriers, we can hope to intercept carriers of atomic bombs.

These are all major problems. They are not substantially different from those which are met in all other branches of the mobilization program. There is a bit of novelty; and there is one important difference: there is a very great lack of military experience. It is doubtful whether the military experience of the end of the last war is relevant, and in any case it applies only to a special form of delivery and a special target, a high altitude delivery of atomic bombs against cities. Nevertheless, I am quite confident that good work on all four of these points is being done, that more and better work will be done, and that, with vigilance and sense, we shall come out with a very considerable increase in military capability.

49

The other side, the policy side, is the one where the role of the public is rather different and rather deeper. That is, of course, also partly a technical question, because one cannot ask whether to use, or under what conditions to use, or how to regard a weapon, until the weapon is defined. It is also a technical question, in that normally and properly these decisions are made by the Chiefs of Staff, by the National Security Council, and by the President, and not by a Gallup Poll. But I think I am right in saying that public opinion on the use of atomic weapons is a most important factor. I have been so assured by many military planners. Even without that assurance, it is obvious, if only because how we use and whether we use atomic weapons in warfare depends a great deal on what else is done. It depends a great deal on whether the public insists upon, supports, or balks at other military or political measures.

The question comes first, of course, in the crude form: Shall we or shall we not use the atomic bomb? I think that before public debate can usefully cope with the question, it is necessary to have a few distinctions. One of them is this: We normally think not of the weapon, but of the specific use which was made of it against Hiroshima and Nagasaki. We think of it as an instrument of strategic bombing, for the destruction of lives and of plants, essentially in cities. It is the decisive, even if perhaps not the final, step in a development that may have started at Guernica, that was characterized by the blitz against London, by the British raids on Hamburg, by our fire raids on Tokyo, and by Hiroshima.

In so far as the prospect of such use may be a deterrent to the initiation of war, or an inducement to governments to carry out policies which we think are sound, and in our interest, it is a fine thing. But the question arises: What happens if the fighting starts? What sort of an instrument is this in a real war? At a time when so very much of our uncommitted military power is in the form of atomic weapons, it is a question that is dangerous not to face. It is not a new question. It has been asked before. I have thought that I could do little better than to quote comment on strategic bombing from the hearings held in October of 1949, before the Armed Services Committee of the House, in connection with the so-called B-36 program. In those hearings, there were many debates about whether the B-36 could ever reach its target, and many debates about whether, if it did, the bombardier could hit the target. From time to time the argument took on a more general character. Here are some fragments of the testimony of Admiral Ralph A. Ofstie, who is now in a Pacific command, who was at that time a member of the Military Liaison Committee to the Atomic Energy Commission.

Admiral Ofstie first said what he meant by strategic bombing. "There is no official definition of the term 'strategic bombing.' The official military term is 'strategic air warfare,' defined as: Air combat and supporting operations designed to effect, through the systematic application of force to a collective series of vital targets, the progressive destruction and disintegration of the enemy's warmaking capacity to a point where he no longer retains the ability or the will to wage war. Vital targets may include key manufacturing systems, sources of raw materials, critical material, stockpiles, power systems, transportation systems, communication facilities, concentrations of uncommitted elements of enemy armed forces, key agricultural areas, and other such target systems.

"This is a very broad field," he said. "Indeed, it would seem to be almost all-inclusive except for the active armed forces of an enemy. In fact, however, the major elements of most of those target systems are located where people live and work, in urban and industrial areas. Further to inject realism into the picture, we must view the tools with which it is proposed the job be done, in this instance the heavy bomber of very long range, of modest performance, operating at great altitudes, and preferably at night. These factors dictate area attack as the means of destroying warmaking capacity located within those areas. Therefore, whether we speak of the mass bombing of World War II or the proposed atomic blitz of today, which are major tenets of the strategic bombing concept, we are talking of attacks on cities. This is what I mean when I use the colloquial term 'strategic bombing.'"

Then, speaking for himself and "many senior officers in the Navy," Admiral Ofstie says: "We consider that strategic air warfare, as practiced in the past and as proposed for the future, is militarily unsound and of limited effect, and is morally wrong, and is decidedly harmful to the stability of a post-war world."

After a technical summary of arguments on the degree of effectiveness, and the technical problems of executing strategic missions, Admiral Ofstie continues: "Much emphasis has been placed upon the instant character of an offensive using atomic bombs. Among laymen this has produced an illusion of power and even a kind of bomb-rattling jingoism. Although responsible officials of the Government generally do not themselves subscribe to it, they must be influenced by the public acceptance of the proposal of instant retaliation. The idea that it is within our power to inflict maximum damage upon the enemy in a short time without serious risk to ourselves creates the delusion that we are stronger than we actually are. This, in turn, becomes a constant temptation for policymakers to overcommit themselves, to make commitments actually impossible to fulfill."

There is nothing in the public record which indicates that these views had at the time any great effect on military or political thought and planning.

This was all long before the fighting broke out in Korea. Much of what was clear to Admiral Ofstie then has become clear to all of us today. The action in Korea, furthermore, has raised publicly another aspect of the question of the use of atomic weapons in warfare: their use against military targets. The targets commonly discussed are troop concentrations, airfields, Naval craft, communications centers. These are among the targets that are an immediate military threat, rather than the basic producing power and the population of an enemy.

I am not qualified, and if I were qualified I would not be allowed, to give a detailed evaluation of the appropriateness of the use of atomic weapons against any or all such targets; but one thing is very clear: It is clear that they can be used only as adjuncts in a military campaign which has some other components, and whose purpose is a military victory. They are not primarily weapons of totality or terror, but weapons used to give combat forces help that they would otherwise lack. They are an integral part of military operations. Only when the atomic bomb is recognized as useful in so far as it is an integral part of military operations, will it really be of much help in the fighting of a war, rather than in warning all mankind to avert it.

Just in this connection, of course, it is clear that the mode of use and the time of use have a relation to each other. Today we do not have very much military strength with which to integrate atomic weapons. Two years from now that should be quite different.

The question of whether to use or not to use atomic weapons is a different question, depending on whether or not one has combat forces and is prepared for combat. They are always terrible weapons; they may not be effective weapons if they are all, or almost all, that we have.

In fact, one can imagine, in some relation to time, at least three different ways in which the atom may serve as military power. The first and the easiest to imagine is as the principal, if not the only, instrument, whose purpose is to destroy plants and kill people: that is the extreme form of the atomic bomb as a strategic weapon.

The second course is the use of atomic bombs primarily against military targets, in tactical use, in coordination with more conventional forms of warfare, in combat. Whether or not they would then be used strategically will depend in part on whether non-use can serve as an effective deterrent; it will depend on the technical advantages, as they appear at the time, of offense and defense. It may not be responsible to anticipate that the strategic use of atomic weapons will be renounced as was the strategic use of gas warfare, because in any future we can foresee, the atomic bomb will offer far vaster prospects of destruction. Such renunciation could, I think, result only from a considered policy decision.

There is a third course we can imagine, that we need to imagine: that, with the obvious horror of a general war, through a combination of our efforts and the efforts of others, and through some good fortune, we may manage to find our way to a more secure and more tolerant and more open world without general war. It is as a principal deterrent to such war that the military power of atomic weapons may yet be decisive.

I am painfully aware that it is not entirely in our hands to determine which of these three courses does, in fact, take place, or which other course. I am also clear that it is not only or primarily a question of the atom bomb. But it is partly a question for the United States, and partly a public question; and it is partly a question of the atom. For if we misjudge what this weapon can or cannot do, in our hands or in the hands of the enemy, if we misjudge its contribution to military strength, it is clear that we will continue to cause our Government, on the basis of our illusions, to follow a course whose only end must be disaster.

About a year ago the Prime Minister of India visited this country. He met with many people and talked with them; and shortly before he left the country I asked him whether he had found in his visit here any appreciation, in this quite different culture, of the Hindu notion of control, of restraint. He answered, "Since this, in the last analysis, only rests on a proper evaluation of the consequences of action, I cannot believe that any great people would be without it."

I believe that the American people are a great people.

With Albert Einstein in LIFE Magazine, 1947. (Photograph by Alfred Eisenstaedt. Courtesy of LIFE Magazine © 1947, Time Inc.)

5

I. UNCOMMON SENSE

1953

A CENTURY AFTER NEWTON, in 1784, the progress of that century was celebrated in an anonymous memorial lodged in the ball of the tower of St. Margaret's church at Gotha, to be found by men of future times. It read:

> "Our days comprise the happiest period of the eighteenth century . . . Hatred born of dogma and the compulsion of conscience sink away; love of man and freedom of thought gain the upper hand. The arts and sciences blossom, and our vision into the workshop of nature goes deep. Artisans approach artists in perfection; useful skills flower at all levels. Here you have a faithful portrait of our time . . . Do the same for those who come after you and rejoice!"

Transience is the backdrop for the play of human progress, for the improvement of man, the growth of his knowledge, the increase of his power, his corruption and his partial redemption. Our civilizations perish; the carved stone, the written word, the heroic act fade into a memory of memory and in the end are gone. The day will come when our race is gone; this house, this earth in which we live will one day be unfit for human habitation, as the sun ages and alters.

Yet no man, be he agnostic or Buddhist or Christian, thinks wholly in these terms. His acts, his thoughts, what he sees of the world around him—the falling of a leaf or a child's joke or the rise of the moon—are part of history; but they are not only part of history; they are a part of becoming and of process but not only that: they partake also of the world outside of time; they partake of the light of eternity.

These two ways of thinking, the way of time and history and the way of eternity and of timelessness, are both part of man's effort to comprehend the world in which he lives. Neither is comprehended in the other nor reducible to it. They are, as we have learned to say in physics, complementary views, each supplementing the other, neither telling the whole story. Let us return to this.

First, we had best review and extend somewhat this account of the complementarity of the physicists. In its simplest form it is that an electron must sometimes be considered as a wave, and sometimes as a particle—a wave, that is, with the continuous propagation and characteristic interference that we learn to understand in the optics laboratory, or as a particle, a thing with well-defined location at any time, discrete and individual and atomic. There is this same duality for all matter and for light. In a little subtler form this complementarity means that there are situations in which the position of an atomic object can be measured and defined and thought about without contradiction; and other situations in which this is not so, but in which other qualities, such as the energy or the impulse of the system, are defined and meaningful. The more nearly appropriate the first way of thinking is to a situation, the more wholly inappropriate the second, so that there are in fact no atomic situations in which both impulse and position will be defined well enough to permit the sort of prediction with which Newtonian mechanics have familiarized us.

It is not only that when we have made an observation on a system and determined, let us say, its position, we do not know its impulse. That is true, but more than that it true. We could say that we know the position of that system and that it may have any one of a number of different impulses. If we try on that basis to predict its behavior as a sort of average behavior of all objects which have the measured position and which have different and unmeasured impulses, and work out the average answer according to Newton's laws, we get a result that is wholly at variance with what we find in nature. This is because of the peculiar property, which has no analogue in the mechanics of large objects, of interference between waves representing the consequences of assuming one impulse and those of assuming another. We are not, that is, allowed to suppose that position and velocity are attributes of an atomic system, some of which we know and others of which we might know but do not. We have to recognize that the attempt to discover these unknown attributes would lose for us the known; that we have a choice, a disjunction; and that this corresponds to the different ways we can go about observing our atom or experimenting with it.

We have a state of affairs completely defined by the nature of the observation and by its outcome—the nature determining what properties of the system will be well defined in the state and what poorly. The outcome then is the determination of the well-defined quantities by measurement. This state thus is a summary, symbolic and uncomfortably abstract for general exposition, of what sort of observation we have made and what we have found through it. It codifies those characteristics of the experimental arrangement which are reliable, in the sense that the equipment we use records something that we know about atomic systems. It describes also those characteristics that are indeterminate, in the sense that they may not only have been disturbed or altered, but that their disturbance cannot be registered or controlled without the loss, in the experiment, of all ability to measure what was supposed to be measured.

This state, this description of the atom, is not the only way of talking about it. It is the only way appropriate to the information we have and the means that we have used to obtain it. It is the full account of this information; and if the experiment was properly and scrupulously done it tells us all that we can find out. It is not all that we could have found out had we chosen a different experiment. It is all that we could find out having chosen this.

This state is objective. We can calculate its properties, reproduce it with similar atoms on another occasion, verify its properties and its ways of change with time. There is no element of the arbitrary or subjective. Once we have done our experiment and its result is recorded and the atom disengaged, we know its meaning and its outcome; we can then forget the details of how we got our information.

But, although the state of the system is objective, a mechanical picture of how it was brought into being is not generally possible. There is a most vivid example of this, made famous by the prominent part it played in the debates between Einstein and Bohr as to the meaning and adequacy of atomic theory. It can be put rather simply. Let us suppose that we have two objects; one of them may be an electron or an atom, and it will be the one we wish to study. The other may be a relatively large piece of matter—a screen with a hole through it, or any other body; but it should be heavy so that its motion will be unimportant compared to that of the electron. Let us suppose that we by measurement know the impulse or momentum of both of these objects, and have them collide. Let the electron go through the hole, or bounce off the other body. If, after the collision, we measure the impulse of the heavy body, we will then know that of the electron because, as Newton's third law teaches us, the sum of the impulses is not altered by the collision. In that case we would have a state of the electron of well-defined impulse, as precisely defined as we had made the precision of our measurements. If, on the other hand, we observed the position of the heavy body, we would know where the light one had been at the moment of the collision, and so would have a quite different description of its state, one in which its position and not its impulse had been well defined—or, in the language of waves, a spherical wave with its center at the point of collision, and not a plane wave with its direction and wave length corresponding to the momentum.

We have thus the option of realizing one or the other of two wholly dissimilar states for the electron, by a choice of what we observe about the heavy body with which it once was in interaction. We are not, in any meaningful sense, physically altering or qualifying the electron; we are defining a part of, although in this case a late part of, the experimental procedure, the very nature of the experiment itself. If we exercise neither option, if we let the heavy body go with unmeasured momentum and undefined position, then we know nothing of the electron at all. It has no state, and we are not prepared to make any meaningful predictions of what will become of it or of what we shall find should we again attempt an experiment upon it The electron cannot be objectified in a manner independent of the means chosen for observing or studying it. The only property we can ascribe to it without such consideration is our total ignorance.

This is a sharp reminder that ways of thinking about things, which seem natural and inevitable and almost appear not to rest on experience so much as on the inherent qualities of thought and nature, do in fact rest on experience; and that there are parts of experience rendered accessible by exploration and experimental refinement where these ways of thought no longer apply.

It is important to remember that, if a very much subtler view of the properties of an electron in an atomic system is necessary to describe the wealth of experience we have had with such systems, it all rests on accepting without revision the traditional accounts of the behavior of large scale objects. The measurements that we have talked about in such highly abstract form do in fact come down in the end to looking at the position of a pointer, or the reading of time on a watch, or measuring out where on a photographic plate or a phosphorescent screen a flash of light or a patch of darkness occurs. They all rest on reducing the experience with atomic systems to experiment and observation made manifest, unambiguous, and objective in the behavior of large objects, where the precautions and incertitudes of the atomic domain no longer directly apply. So it is that ever-increasing refinements and critical revisions in the way we talk about remote or small or inaccessible parts of the physical world have no direct relevance to the familiar physical world of common experience.

Common sense is not wrong in the view that it is meaningful, appropriate, and necessary to talk about the large objects of our daily experience as though they had a velocity that we knew, and a place that we knew, and all the rest of it. Common sense is wrong only if it insists that what is familiar must reappear in what is unfamiliar. It is wrong only if it leads us to expect that every country the we visit is like the last country we saw. Common sense, as the common heritage from the millennia of common life, may lead us into error if we wholly forget the circumstances to which that common life has been restricted.

Misunderstanding of these relations has led men to wish to draw from new discoveries, and particularly those in the atomic domain, far-reaching consequences for the ordinary affairs of men. Thus it was noted that, since the ultimate laws of atomic behavior are not strictly causal, not strictly determinate, the famous argument of Laplace for a wholly determinate universe could not be maintained. And there were men who believed that they had discovered in the acausal and indeterminate character of atomic events the physical basis for that sense of freedom which characterizes man's behavior in the face of decision and responsibility.

In a similar lighthearted way it was pointed out that, as the state of an atomic system requires observation for its definition, so the course of psychological phenomena might be irretrievably altered by the very effort to probe them—as a man's thoughts are altered by the fact that he has formulated and spoken them. It is, of course, not the fact that observation may change the state of an atomic system that gives rise to the need for a complementary description; it is the fact that, if the observation is to be meaningful, it will preclude any analysis or control of that change, that is decisive.

But these misapplications of the findings of atomic physics to human affairs do not establish that there are no valid analogies. These analogies will, in the nature of things, be less sharp, less compelling, less ingenious. They will rest upon the fact that complementary modes of thought and complementary descriptions of reality are an old, long-enduring part of our tradition. All that the experience of atomic physics can do in these affairs is to give us a reminder, and a certain reassurance, that these ways of talking and thinking can be factual, appropriate, precise, and free of obscurantism.

There are a number of examples which are illuminated by, and in turn illuminate, the complementarity of atomic theory. Some of them are from quite different parts of human life and some of them from older parts of science. There is one from physics itself which is revealing, both in its analogies and its points of difference. One of the great triumphs of nineteenth-century physics was the kinetic theory of heat—what is called statistical mechanics. This is both an interpretation and a deduction of many of the large-scale properties and tendencies of matter: of the tendency, for instance, of bodies that can exchange heat to come to a common temperature, or of the density of a gas to be uniform throughout a container, or of work to dissipate itself in heat, or quite generally of all of those irreversible processes in nature wherein the entropy of systems increases, and forms become more uniform and less differentiated when left to themselves to develop.

The phenomena we deal with here are defined in terms of temperature and density and pressure and other large-scale properties. The kinetic theory, statistical mechanics, interprets the behavior of these systems in terms of the forces acting on the molecules and of the motion of the molecules that compose them, which are usually quite accurately described by Newton's laws. But it is a statistical theory of this motion, recognizing that in fact we do not in general know, and are not in detail concerned with, the positions and velocities of the molecules themselves, but only with their average behavior. We interpret the temperature of a gas, for instance, in terms of the average kinetic energy of its molecules, and the pressure as the average of the forces exerted by the collision of these molecules on the surface of the container. This description in terms of averages, embodying as part of itself our ignorance of the detailed state of affairs, is thus in some sense complementary to a complete dynamic description in terms of the motion of the individual molecules. In this sense kinetic theory and dynamics are complementary. One applies to a situation in which the individual patterns of molecular behavior are known and studied; the other applies to a situation largely defined by our ignorance of these patterns.

But the analogy to atomic complementarity is only partial, because there is nothing in the classical dynamics which underlies kinetic theory to suggest that the behavior of a gas would be any different if we had performed the immense job of locating and measuring what all the molecules were doing. We might then, it is true, not find it natural to talk about temperature, because we would need no average behavior; we would have an actual one; but we could still define the temperature in terms of the total kinetic energy of the molecules, and we would still find that it tended to equalize between one part of the system and another.

We have therefore a situation in which there are two ways of describing a system, two sets of concepts, two centers of pre-occupation. One is appropriate when we are dealing with a very few molecules and want to know what those molecules do; the other appropriate when we have a large mass of matter and only rough and large-scale observations about it.

There is, however, no logical or inherent difficulty within the framework of classical physics, in combining both descriptions for a single system—and classical physics, we repeat, is adequate for most, if not all, of these problems of statistical mechanics. It is not that we cannot do this without violating the laws of physics; it is that it makes no sense to do it, since each description is appropriate to a context quite different from the other. It is clear that, if we insisted on the detailed description of the motion of individual molecules, the notions of probability which turn out to be so essential for our understanding of the irreversible character of physical events in nature would never enter. We should not have the great insight that we now do: namely, that the direction of change in the world is from the less probable to the more, from the more organized to the less, because all we would be talking about would be an incredible number of orbits and trajectories and collisions. It would be a great miracle to us that, out of equations of motion, which to every allowed motion permit a precisely opposite one, we could nevertheless emerge into a world in which there is a trend of change with time which is irreversible, unmistakable, and familiar in all our physical experience.

In considering the relations between the various sciences, there are similar instances of complementary views. In many cases, it is not clear whether this is the sort of complementarity that we have between the statistical and dynamic descriptions of a gas, a contrast of interest and terminology, but not an inherent inapplicability of two ways of talking; or whether on the contrary the situation is in fact more as it is in atomic physics, where the nature of the world is such that the two modes of description cannot be applied at once to the same situation. Every science has its own language. But dictionaries of translation between the languages do exist, and mark an ever-growing understanding and unity of science as a whole. It is not always clear whether the dictionaries will be complete; between physics and chemistry they apparently are. Everything the chemist observes and describes can be talked about in terms of atomic mechanics, and most of it at least can be understood. Yet no one suggests that, in dealing with the complex chemical forms which are of biological interest, the language of atomic physics would be helpful. Rather it would tend to obscure the great regularities of biochemistry, as the dynamic description of a gas would obscure its thermodynamic behavior.

The contrast becomes even more marked when we consider the physico-chemical description of living forms. Here, in spite of the miraculous sharpness of the tools of chemical analysis, of the extensive use not only of the microscope but of the electron microscope to determine fine details of biological structure, in spite of the use of tracers to follow changes on a molecular scale, questions have still been raised as to whether this description can in the nature of things be complete.

The question involves two points: the first having to do with the impossibility of wholly isolating a biological system from its physical environment without killing it; the second with the possibility that a really complete physico-chemical study of the pivotal structures in biological processes—of genes, let us say, in the nuclei of dividing cells—might not be compatible with the undisturbed course of life itself. It would appear to be the general opinion of biologists that no such limitations will prove decisive; that a complete description of biology will be possible not only in terms of the concepts of biology but in terms reducible to those of physics and chemistry. Certainly it is a large part of the aim and wonder of biological progress to carry this program as far as possible.

Analogous questions appear much sharper, and their answer more uncertain, when we think of the phenomena of consciousness; and, despite all the progress that has been made in the physiology of the sense organs and of the brain, despite our increasing knowledge of these intricate marvels both as to their structure and their functioning, it seems rather unlikely that we shall be able to describe in physico-chemical terms the physiological phenomena which accompany a conscious thought, or sentiment, or will. Today the outcome is uncertain. Whatever the outcome, we know that, should an understanding of the physical correlate of elements of consciousness indeed be available, it will not itself be the appropriate description for the thinking man himself, for the clarification of his thoughts, the resolution of his will, or the delight of his eye and mind at works of beauty. Indeed, an understanding of the complementary nature of conscious life and its physical interpretation appears to me a lasting element in human understanding and a proper formulation of the historic views called psycho-physical parallelism.

For within conscious life, and in its relations with the description of the physical world, there are again many examples. There is the relation between the cognitive and the affective sides of our lives, between knowledge or analysis and emotion or feeling. There is the relation between the aesthetic and the heroic, between feeling and that precursor and definer of action, the ethical commitment; there is the classical relation between the analysis of one's self, the determination of one's motives and purposes, and that freedom of choice, that freedom of decision and action, which are complementary to it.

Whether a physico-chemical description of the material counterpart of consciousness will in fact ever be possible, whether physiological or psychological observation will ever permit with any relevant confidence the prediction of our behavior in moments of decision and in moments of challenge, we may be sure that these analyses and these understandings, even should they exist, will be as irrelevant to the acts of decision and the castings of the will as are the trajectories of molecules to the entropy of a gas. To be touched with awe, or humor, to be moved by beauty, to make a commitment or a determination, to understand some truth—these are complementary modes of the human spirit. All of them are part of man's spiritual life. None can replace the others, and where one is called for the others are in abeyance.

Just as with the α-particles of Rutherford, which were first for him an object of study and then became for him a tool of study, a tool for investigating other objects, so our thoughts and words can be the subject of reflection and analysis; so we can be introspective, critical, and full of doubt. And so, in other times and other contexts, these same words, these same thoughts taken as instruments, are the power of human understanding itself, and the means of our further enlightenment.

The wealth and variety of physics itself, the greater wealth and variety of the natural sciences taken as a whole, the more familiar, yet still strange and far wider wealth of the life of the human spirit, enriched by complementary, not at once compatible ways, irreducible one to the other, have a greater harmony. They are the elements of man's sorrow and his splendor, his frailty and his power, his death, his passing, and his undying deeds.

II. AN OPEN HOUSE

1953

[...]WE HAVE LOOKED TOGETHER into one of the rooms of the house called "science." This is a relatively quiet room that we know as quantum theory or atomic theory. The great girders which frame it, the lights and shadows and vast windows—these were the work of a generation of our predecessors more than two decades ago. It is not wholly quiet. Young people visit it and study in it and pass on to other chambers; and from time to time someone rearranges a piece of the furniture to make the whole more harmonious; and many, as we have done, peer through its windows or walk through it as sightseers. It is not so old but that one can hear the sound of the new wings being built nearby, where men walk high in the air to erect new scaffoldings, not unconscious of how far they may fall. All about there are busy workshops where the builders are active, and very near indeed are those of us who, learning more of the primordial structure of matter, hope some day for chambers as fair and lovely as that in which we have spent the years of our youth and our prime.

It is a vast house indeed. It does not appear to have been built upon any plan but to have grown as a great city grows. There is no central chamber, no one corridor from which all others debouch. All about the periphery men are at work studying the vast reaches of space and the state of affairs billions of years ago; studying the intricate and subtle but wonderfully meet mechanisms by which life proliferates, alters, and endures; studying the reach of the mind and its ways of learning; digging deep into the atoms and the atoms within atoms and their unfathomed order. It is a house so vast that none of us know it, and even the most fortunate have seen most rooms only from the outside or by a fleeting passage, as in a king's palace open to visitors. It is a house so vast that there is not and need not be complete concurrence on where its chambers stop and those of the neighboring mansions begin.

It is not arranged in a line nor a square nor a circle nor a pyramid, but with a wonderful randomness suggestive of unending growth and improvisation. Not many people live in the house, relatively speaking—perhaps if we count all its chambers and take residence requirements quite lightly, one tenth of one per cent, of all the people in this world—probably, by any reasonable definition, far fewer. And even those who live here live elsewhere also, live in houses where the rooms are not labelled atomic theory or genetics or the internal constitution of the stars, but quite different names like power and production and evil and beauty and history and children and the word of God.

We go in and out; even the most assiduous of us is not bound to this vast structure. One thing we find throughout the house: there are no locks; there are no shut doors; wherever we go there are the signs and usually the words of welcome. It is an open house, open to all comers.

The discoveries of science, the new rooms in this great house, have changed the way men think of things outside its walls. We have some glimmering now of the depth in time and the vastness in space of the physical world we live in. An awareness of how long our history and how immense our cosmos touches us even in simple earthly deliberations. We have learned from the natural history of the earth and from the story of evolution to have a sense of history, of time and change. We learn to talk of ourselves, and of the nature of the world and its reality as not wholly fixed in a silent quiet moment, but as unfolding with novelty and alteration, decay and new growth. We have understood something of the inner harmony and beauty of strange primitive cultures, and through this see the qualities of our own life in an altered perspective, and recognize its accidents as well as its inherent necessities. We are, I should think, not patriots less but patriots very differently for loving what is ours and understanding a little of the love of others for their lands and ways. We have begun to understand that it is not only in his rational life that man's psyche is intelligible, that even in what may appear to be his least rational actions and sentiments we may discover a new order. We have the beginnings of an understanding of what it is in man, and more in simple organisms, that is truly heritable, and rudimentary clues as to how the inheritance occurs. We know, in surprising detail, what is the physical counterpart of the act of vision and of other modes of perception. Not one of these new ideas and new insights is so little, or has so short a reach in its bearing on the common understanding but that it alone could make a proper theme for "Science and the Common Understanding." Yet we have been, bearing in mind my limited area of experience, in that one room of the part of the house where physics is, in which I have for some years worked and taught.

In that one room—in that relatively quiet room where we have been together—we have found things quite strange for those who have not been there before, yet reminiscent of what we have seen in other houses and known in other days. We have seen that in the atomic world we have been led by experience to use descriptions and ideas that apply to the large-scale world of matter, to the familiar world of our schoolday physics; ideas like the position of a body and its acceleration and its impulse and the forces acting on it; ideas like wave and interference; ideas like cause and probability. But what is new, what was not anticipated a half-century ago, is that, though to an atomic system there is a potential applicability of one or another of these ideas, in any real situation only some of these ways of description can be actual. This is because we need to take into account not merely the atomic system we are studying, but the means we use in observing it, and the fitness of these experimental means for defining and measuring selected properties of the system. All such ways of observing are needed for the whole experience of the atomic world; all but one are excluded in any actual experience. In the specific instance, there is a proper and consistent way to describe what the experience is; what it implies; what it predicts and thus how to deal with its consequences. But any such specific instance excludes by its existence the application of other ideas, other modes of prediction, other consequences. They are, we say, complementary to one another; atomic theory is in part an account of these descriptions and in part an understanding of the circumstances to which one applies, or another or another.

And so it is with man's life. He may be any of a number of things; he will not be all of them. He may be well versed, he may be a poet, he may be a creator in one or more than one science; he will not be all kinds of man or all kinds of scientist; and he will be lucky if he has a bit of familiarity outside the room in which he works.

So it is with the great antinomies that through the ages have organized and yet disunited man's experience: the antinomy between the ceaseless change and wonderful novelty and the perishing of all earthly things, and the eternity which inheres in every happening; in the antinomy between growth and order, between the spontaneous and changing and irregular and the symmetrical and balanced; in the related antinomy between freedom and necessity; between action, the life of the will, and observation and analysis and the life of reason; between the question "how?" and the questions "why?" and "to what end?"; between the causes that derive from natural law, from unvarying regularities in the natural world, and those other causes that express purposes and define goals and ends.

So it is in the antinomy between the individual and the community; man who is an end in himself and man whose tradition, whose culture, whose works, whose words have meaning in terms of other men and his relations to them. All our experience has shown that we can neither think, nor in any true sense live, without reference to these antinomic modes. We cannot in any sense be both the observers and the actors in any specific instance, or we shall fail properly to be either one or the other; yet we know that our life is built of these two modes, is part free and part inevitable, is part creation and part discipline, is part acceptance and part effort. We have no written rules that assign us to these ways; but we know that only folly and death of the spirit result when we deny one or the other, when we erect one as total and absolute and make the others derivative and secondary. We recognize this when we live as men. We talk to one another; we philosophize; we admire great men and their moments of greatness; we read; we study; we recognize and love in a particular act that happy union of the generally incompatible. With all of this we learn to use some reasonable part of the full register of man's resources.

We are, of course, an ignorant lot; even the best of us knows how to do only a very few things well; and of what is available in knowledge of fact, whether of science or of history, only the smallest part is in any one man's knowing.

The greatest of the changes that science has brought is the acuity of change; the greatest novelty the extent of novelty. Short of rare times of great disaster, civilizations have not known such rapid alteration in the conditions of their life, such rapid flowering of many varied sciences, such rapid changes in the ideas we have about the world and one another. What has been true in the days of a great disaster or great military defeat for one people at one time is true for all of us now, in the sense that our ends have little in common with our beginnings. Within a lifetime what we learned at school has been rendered inadequate by new discoveries and new inventions; the ways that we learn in childhood are only very meagerly adequate to the issues that we must meet in maturity.

In fact, of course, the notion of universal knowledge has always been an illusion; but it is an illusion fostered by the monistic view of the world in which a few great central truths determine in all its wonderful and amazing proliferation everything else that is true. We are not today tempted to search for these keys that unlock the whole of human knowledge and of man's experience. We know that we are ignorant; we are well taught it, and the more surely and deeply we know our own job the better able we are to appreciate the full measure of our pervasive ignorance. We know that these are inherent limits, compounded, no doubt, and exaggerated by that sloth and that complacency without which we would not be men at all.

But knowledge rests on knowledge; what is new is meaningful because it departs slightly from what was known before; this is a world of frontiers, where even the liveliest of actors or observers will be absent most of the time from most of them. Perhaps this sense was not so sharp in the village—that village which we have learned a little about but probably do not understand too well— the village of slow change and isolation and fixed culture which evokes our nostalgia even if not our full comprehension. Perhaps in the villages men were not so lonely; perhaps they found in each other a fixed community, a fixed and only slowly growing store of knowledge—a single world. Even that we may doubt, for there seem to be always in the culture of such times and places vast domains of mystery, if not unknowable, then imperfectly known, endless and open.

As for ourselves in these times of change, of ever-increasing knowledge, of collective power and individual impotence, of heroism and of drudgery, of progress and of tragedy, we too are brothers. And if we, who are the inheritors of two millennia of Christian tradition, understand that for us we have come to be brothers second by being children first, we know that in vast parts of the world where there has been no Christian tradition, and with men who never have been and never may be Christian in faith there is nevertheless a bond of brotherhood. We know this not only because of the almost universal ideal of human brotherhood and human community; we know it at first hand from the more modest, more diverse, more fleeting associations which are the substance of our life. The ideal of brotherhood, the ideal of fraternity in which all men, wicked and virtuous, wretched and fortunate, are banded together has its counterpart in the experience of communities, not ideal, not universal, imperfect, impermanent, as different from the ideal and as reminiscent of it as are the ramified branches of science from the ideal of a unitary, all-encompassing science of the eighteenth century.

Each of us knows from his own life how much even a casual and limited association of men goes beyond him in knowledge, in understanding, in humanity, and in power. Each of us, from a friend or a book or by concerting of the little we know with what others know, has broken the iron circle of his frustration. Each of us has asked help and been given it, and within our measure each of us has offered it. Each of us knows the great new freedom sensed almost as a miracle, that men banded together for some finite purpose experience from the power of their common effort. We are likely to remember the times of the last war, where the common danger brought forth in soldier, in worker, in scientist, and engineer a host of new experiences of the power and the comfort in even bleak undertakings, of common, concerted, cooperative life. Each of us knows how much he has been transcended by the group of which he has been or is a part; each of us has felt the solace of other men's knowledge to stay his own ignorance, of other men's wisdom to stay his folly, of other men's courage to answer his doubts or his weakness.

These are the fluid communities, some of long duration when circumstances favored—like the political party or many a trade union—some fleeting and vivid, encompassing in the time of their duration a moment only of the member's life; and in our world at least they are ramified and improvised, living and dying, growing and falling off almost as a form of life itself. This may be more true of the United States than of any other country. Certainly the bizarre and comical aspects impressed de Tocqueville more than a century ago when he visited our land and commented on the readiness with which men would band together: to improve the planting of a town, or for political reform, or for the pursuit or interexchange of knowledge, or just for the sake of banding together, because they liked one another or disliked someone else. Circumstances may have exaggerated the role of the societies, of the fluid and yet intense communities in the United States; yet these form a common pattern for our civilization. It brought men together in the Royal Society and in the French Academy and in the Philosophical Society that Franklin founded, in family, in platoon, on a ship, in the laboratory, in almost everything but a really proper club.

If we err today—and I think we do—it is in expecting too much of knowledge from the individual and too much of synthesis from the community. We tend to think of these communities, no less than of the larger brotherhood of man, as made up of individuals, as composed of them as an atom of its ingredients. We think similarly of general laws and broad ideas as made up of the instances which illustrate them, and from an observation of which we may have learned them.

Yet this is not the whole. The individual event, the act, goes far beyond the general law. It is a sort of intersection of many generalities, harmonizing them in one instance as they cannot be harmonized in general. And we as men are not only the ingredients of our communities; we are their intersection, making a harmony which does not exist between the communities except as we, the individual men, may create it and reveal it. So much of what we think, our acts, our judgments of beauty and of right and wrong, come to us from our fellow men that what would be left

73

were we to take all this away would be neither recognizable nor human. We are men because we are part of, but not because only part of, communities; and the attempt to understand man's brotherhood in terms only of the individual man is as little likely to describe our world as is the attempt to describe general laws as the summary of their instances. These are indeed two complementary views, neither reducible to the other, no more reducible than is the electron as wave to the electron as particle.

And this is the mitigant of our ignorance. It is true that none of us will know very much; and most of us will see the end of our days without understanding in all its detail and beauty the wonders uncovered even in a single branch of a single science. Most of us will not even know, as a member of any intimate circle, anyone who has such knowledge; but it is also true that, although we are sure not to know everything and rather likely not to know very much, we can know anything that is known to man, and may, with luck and sweat, even find out some things that have not before been known to him. This possibility, which, as a universal condition of man's life is new, represents today a high and determined hope, not yet a reality; it is for us in England and in the United States not wholly remote or unfamiliar. It is one of the manifestations of our belief in equality, that belief which could perhaps better be described as a commitment to unparalleled diversity and unevenness in the distribution of attainments, knowledge, talent, and power.

This open access to knowledge, these unlocked doors and signs of welcome, are a mark of a freedom as fundamental as any. They give a freedom to resolve difference by converse, and, where converse does not unite, to let tolerance compose diversity. This would appear to be a freedom barely compatible with modern political tyranny. The multitude of communities, the free association for converse or for common purpose, are acts of creation. It is not merely that without them the individual is the poorer; without them a part of human life, not more nor less fundamental than the individual, is foreclosed. It is a cruel and humorless sort of pun that so powerful a present form of modern tyranny should call itself by the very name of a belief in community, by a word "communism" which in other times evoked memories of villages and village inns and of artisans concerting their skills, and of men of learning content with anonymity. But perhaps only a malignant end can follow the systematic belief that all communities are one community; that all truth is one truth; that all experience is compatible with all other; that total knowledge is possible; that all that is potential can exist as actual. This is not man's fate; this is not his path; to force him on it makes him resemble not that divine image of the all-knowing and all-powerful but the helpless, iron-bound prisoner of a dying world. The open society, the unrestricted access to knowledge, the unplanned and uninhibited association of men for its furtherance—these are what may make a vast, complex, ever-growing, ever-changing, ever more specialized and expert technological world nevertheless a world of human community.

So it is with the unity of science—that unity that is far more a unity of comparable dedication than a unity of common total understanding. This heartening phrase, "the unity of science," often tends to evoke a wholly false picture, a picture of a few basic truths, a few critical techniques, methods, and ideas, from which all discoveries and understanding of science derive; a sort of central exchange, access to which will illuminate the atoms and the galaxies, the genes and the sense organs. The unity of science is based rather on just such a community as I have described. All parts of it are open to all of us, and this is no merely formal invitation. The history of science is rich in example of the fruitfulness of bringing two sets of techniques, two sets of ideas, developed in separate contexts for the pursuit of new truth, into touch with one another. The sciences fertilize each other; they grow by contact and by common enterprise. Once again, this means that the scientist may profit from learning about any other science; it does not mean that he must learn about them all. It means that the unity is a potential unity, the unity of the things that might be brought together and might throw light one on the other. It is not global or total or hierarchical.

Even in science, and even without visiting the room in its house called atomic theory, we are again and again reminded of the complementary traits in our own life, even in our own professional life. We are nothing without the work of others our predecessors, others our teachers, others our contemporaries. Even when, in the measure of our adequacy and our fullness, new insight and new order are created, we are still nothing without others. Yet we are more.

There is a similar duality in our relations to wider society. For society our work means many things: pleasure, we hope, for those who follow it; instruction for those who perhaps need it; but also and far more widely, it means a common power, a power to achieve that which could not be achieved without knowledge. It means the cure of illness and the alleviation of suffering; it means the easing of labor and the widening of the readily accessible frontiers of experience, of communication, and of instruction. It means, in an earthy way, the power of betterment—that riddled word. We are today anxiously aware that the power to change is not always necessarily good.

As new instruments of war, of newly massive terror, add to the ferocity and totality of warfare, we understand that it is a special mark and problem of our age that man's ever-present preoccupation with improving his lot, with alleviating hunger and poverty and exploitation, must be brought into harmony with the overriding need to limit and largely to eliminate resort to organized violence between nation and nation. The increasingly expert destruction of man's spirit by the power of police, more wicked if not more awful than the ravages of nature's own hand, is another such power, good only if never to be used.

We regard it as proper and just that the patronage of science by society is in large measure based on the increased power which knowledge gives. If we are anxious that the power so given and so obtained be used with wisdom and with love of humanity, that is an anxiety we share with almost everyone. But we also know how little of the deep new knowledge which has altered the face of the world, which has changed—and increasingly and ever more profoundly must change—man's views of the world, resulted from a quest for practical ends or an interest in exercising the power that knowledge gives. For most of us, in most of those moments when we were most free of corruption, it has been the beauty of the world of nature and the strange and compelling harmony of its order, that has sustained, inspirited, and led us. That also is as it should be. And if the forms in which society provides and exercises its patronage leave these incentives strong and secure, new knowledge will never stop as long as there are men.

We know that our work is rightly both an instrument and an end. A great discovery is a thing of beauty; and our faith—our binding, quiet faith—is that knowledge is good and good in itself. It is also an instrument; it is an instrument for our successors, who will use it to probe elsewhere and more deeply; it is an instrument for technology, for the practical arts, and for man's affairs. So it is with us as scientists; so it is with us as men. We are at once instrument and end, discoverers and teachers, actors and observers. We understand, as we hope others understand, that in this there is a harmony between knowledge in the sense of science, that specialized and general knowledge which it is our purpose to uncover, and the community of man. We, like all men, are among those who bring a little light to the vast unending darkness of man's life and world. For us as for all men, change and eternity, specialization and unity, instrument and final purpose, community and individual man alone, complementary each to the other, both require and define our bonds and our freedom.

At Los Alamos in 1946. First row from the left: N. Bradbury, J. Manley, and Enrico Fermi. Robert Oppenheimer is seated in the second row. (LAL)

6

PROSPECTS IN THE ARTS AND SCIENCES

1954

THE WORDS "prospects in the arts and sciences" mean two quite different things to me. One is prophecy: What will the scientists discover and the painters paint, what new forms will alter music, what parts of experience will newly yield to objective description? The other meaning is that of a view: What do we see when we look at the world today and compare it with the past? I am not a prophet; and I cannot very well speak to the first subject, though in many ways I should like to. I shall try to speak to the second, because there are some features of this view which seem to me so remarkable, so new and so arresting, that it may be worth turning our eyes to them; it may even help us to create and shape the future better, though we cannot foretell it.

In the arts and in the sciences, it would be good to be a prophet. It would be a delight to know the future. I had thought for a while of my own field of physics and of those nearest to it in the natural sciences. It would not be too hard to outline the questions which natural scientists today are asking themselves and trying to answer. What, we ask in physics, is matter, what is it made of, how does it behave when it is more and more violently atomized, when we try to pound out of the stuff around us the ingredients which only violence creates and makes manifest? What, the chemists ask, are those special features of nucleic acids and proteins which make life possible and give it its characteristic endurance and mutability? What subtle chemistry, what arrangements, what reactions and controls make the cells of living organisms differentiate so that they may perform functions as oddly diverse as

79

transmitting information throughout our nervous systems or covering our heads with hair? What happens in the brain to make a record of the past, to hide it from consciousness, to make it accessible to recall? What are the physical features which make consciousness possible?

All history teaches us that these questions that we think the pressing ones will be transmuted before they are answered, that they will be replaced by others, and that the very process of discovery will shatter the concepts that we today use to describe our puzzlement.

It is true that there are some who profess to see in matters of culture, in matters precisely of the arts and sciences, a certain macrohistorical pattern, a grand system of laws which determines the course of civilization and gives a kind of inevitable quality to the unfolding of the future. They would, for instance, see the radical, formal experimentation which characterized the music of the last half-century as an inevitable consequence of the immense flowering and enrichment of natural science; they would see a necessary order in the fact that innovation in music precedes that in painting and that in turn in poetry, and point to this sequence in older cultures. They would attribute the formal experimentation of the arts to the dissolution, in an industrial and technical society, of authority—of secular, political authority, and of the catholic authority of the church. They are thus armed to predict the future. (But this, I fear, is not my dish.)

If a prospect is not a prophecy, it is a view. What does the world of the arts and sciences look like? There are two ways of looking at it: One is the view of the traveler, going by horse or foot, from village to village to town, staying in each to talk with those who live there and to gather something of the quality of its life. This is the intimate view, partial, somewhat accidental, limited by the limited life and strength and curiosity of the traveler, but intimate and human, in a human compass. The other is the vast view, showing the earth with its fields and towns and valleys as they appear to a camera carried in a high-altitude rocket. In one sense this prospect will be more complete; one will see all branches of knowledge, one will see all the arts, one will see them as part of the vastness and complication of the whole of human life on earth. But one will miss a great deal; the beauty and warmth of human life will largely be gone from that prospect.

It is in this vast high-altitude survey that one sees the general surprising quantitative features that distinguish our time. This is where the listings of science and endowments and laboratories and books published show up; this is where we learn that more people are engaged in scientific research today than ever before, that the Soviet world and the free world are running neck and neck in the training of scientists, that more books are published per capita in England than in the United States, that the social sciences are pursued actively in America, Scandinavia, and England, that there are more people who hear the great music of the past, and more music composed and more paintings painted. This is where we learn that the arts and sciences are flourishing. This great map, showing the world from afar and almost as to a stranger, would show more: It would show the immense diversity of culture and life, diversity in place and tradition for the first time clearly manifest on a world-wide scale, diversity in technique and language, separating science from science and art from art, and all of one from all of the other. This great map, world-wide, culture-wide, remote, has some odd features. There are innumerable villages. Between the villages there appear to be almost no paths discernible from this high altitude. Here and there passing near a village, sometimes through its heart, there will be a superhighway, along which windy traffic moves at enormous speed. The superhighways seem to have little connection with villages, starting anywhere, ending anywhere, and sometimes appearing almost by design to disrupt the quiet of the village. This view gives us no sense of order or of unity. To find these we must visit the villages, the quiet, busy places, the laboratories and studies and studios. We must see the paths that are barely discernible; we must understand the superhighways and their dangers.

In the natural sciences these are and have been and are likely to continue to be heroic days. Discovery follows discovery, each both raising and answering questions, each ending a long search, and each providing the new instruments for a new search. There are radical ways of thinking unfamiliar to common sense and connected with it by decades or centuries of increasingly special-ized and unfamiliar experience. There are lessons of how limited, for all its variety, the common experience of man has been with regard to natural phenomena, and hints and analogies as to how limited may be his experience with man. Every new finding is a part of the instrument kit of the sciences for further investigation and for penetrating into new fields. Discoveries of knowledge fructify technology and the practical arts, and these in turn pay back refined techniques, new possibilities of observation and experiment.

In any science there is harmony between practitioners. A man may work as an individual, learning of what his colleagues do through reading or conversation; he may be working as a member of a group on problems whose technical equipment is too massive for individual effort. But whether he is a part of a team or solitary in his own study, he, as a professional, is a member of a community. His colleagues in his own branch of science will be grateful to him for the inventive or creative thoughts he has, will welcome his criticism. His world and work will be objectively communicable; and he will be quite sure that if there is error in it, that error will not long be undetected. In his own line of work he lives in a community where common understanding combines with common purpose and interest to bind men together both in freedom and in co-operation.

This experience will make him acutely aware of how limited, how inadequate, how precious is this condition of his life; for in his relations with a wider society, there will be neither the sense of community nor of objective understanding. He will sometimes find, in returning to practical undertakings, some sense of community with men who are not expert in his science, with other scientists whose work is remote from his, and with men of action and men of art. The frontiers of science are separated now by long years of study, by specialized vocabularies, arts, techniques, and knowledge from the common heritage even of a most civilized society; and anyone working at the frontier of such science is in that sense a very long way from home, a long way too from the practical arts that were its matrix and origin, as indeed they were of what we today call art.

The specialization of science is an inevitable accompaniment of progress; yet it is full of dangers, and it is cruelly wasteful, since so much that is beautiful and enlightening is cut off from most of the world. Thus it is proper to the role of the scientist that he not merely find new truth and communicate it to his fellows, but that he teach, that he try to bring the most honest and intelligible account of new knowledge to all who will try to learn. This is one reason—it is the decisive organic reason—why scientists belong in universities. It is one reason why the patronage of science by and through universities is its most proper form; for it is here, in teaching, in the association of scholars and in the friendships of teachers and taught, of men who by profession must themselves be both teachers and taught, that the narrowness of scientific life can best be moderated, and that the analogies, insights, and harmonies of scientific discovery can find their way into the wider life of man.

In the situation of the artist today there are both analogies to and differences from that of the scientist; but it is the differences which are the most striking and which raise the problems that touch most on the evil of our day. For the artist it is not enough that he communicate with others who are expert in his own art. Their fellowship, their understanding, and their appreciation may encourage him; but that is not the end of his work, nor its nature. The artist depends on a common sensibility and culture, on a common meaning of symbols, on a community of experience and common ways of describing and interpreting it. He need not write for everyone or paint or play for everyone. But his audience must be man; it must be man, and not a specialized set of experts among his fellows. Today that is very difficult. Often the artist has an aching sense of great loneliness, for the community to which he addresses himself is largely not there; the traditions and the culture, the symbols and the history, the myths and the common experience, which it is his function to illuminate, to harmonize, and to portray, have been dissolved in a changing world.

There is, it is true, an artificial audience maintained to moderate between the artist and the world for which he works: the audience of the professional critics, popularizers, and advertisers of art. But though, as does the popularizer and promoter of science, the critic fulfills a necessary present function and introduces some order and some communication between the artist and the world, he cannot add to the intimacy and the directness and the depth with which the artist addresses his fellow men.

To the artist's loneliness there is a complementary great and terrible barrenness in the lives of men. They are deprived of the illumination, the light and tenderness and insight of an intelligible interpretation, in contemporary terms, of the sorrows and wonders and gaieties and follies of man's life. This may be in part offset, and is, by the great growth of techical means for making the art of the past available. But these provide a record of past intimacies between art and life; even when they are applied to the writing and painting and composing of the day, they do not bridge the gulf between a society, too vast and too disordered, and the artist trying to give meaning and beauty to its parts.

In an important sense this world of ours is a new world, in which the unity of knowledge, the nature of human communities, the order of society, the order of ideas, the very notions of society and culture have changed and will not return to what they have been in the past. What is new is new not because it has never been there before, but because it has changed in quality. One thing that is new is the prevalence of newness, the changing scale and scope of change itself, so that the world alters as we walk in it, so that the years of man's life measure not some small growth or rearrangement or moderation of what he learned in childhood, but a great upheaval. What is new is that in one generation our knowledge of the natural world engulfs, upsets, and complements all knowledge of the natural world before. The techniques, among which and by which we live, multiply and ramify, so that the whole world is bound together by communication, blocked here and there by the immense synapses of political tyranny. The global quality of the world is new: our knowledge of and sympathy with remote and diverse peoples, our involvement with them in practical terms, and our commitment to them in terms of brotherhood. What is new in the world is the massive character of the dissolution and corruption of authority, in belief, in ritual, and in temporal order. Yet this is the world that we have come to live in. The very difficulties which it presents derive from growth in understanding, in skill, in power. To assail the changes that have unmoored us from the past is futile, and in a deep sense, I think, it is wicked. We need to recognize the change and learn what resources we have.

Again I will turn to the schools and, as their end and as their center, the universities. For the problem of the scientist is in this respect not different from that of the artist or of the historian. He needs to be a part of the community, and the community can only with loss and peril be without him. Thus it is with a sense of interest and hope that we see a growing recognition that the creative artist is a proper charge on the university, and the university a proper home for him; that a composer or a poet or a playwright or painter needs the toleration, understanding, the rather local and parochial patronage that a university can give; and that this will protect him from the tyranny of man's communication and professional promotion. For here there is an honest chance that what the artist has of insight and of beauty will take root in the community, and that some intimacy and some human bonds can mark his relations with his patrons. For a university rightly and inherently is a place where the individual man can form new syntheses, where the accidents of friendship and association can open a man's eyes to a part of science or art which he had not known before, where parts of human life, remote and perhaps superficially incompatible, can find in men their harmony and their synthesis.

These, then, in rough and far too general words, are some of the things we see as we walk through the villages of the arts and of the sciences and notice how thin are the paths that lead from one to another, and how little in terms of human understanding and pleasure the work of the villages comes to be shared outside.

The superhighways do not help. They are the mass media— from the loudspeakers in the deserts of Asia Minor and the cities of Communist China to the organized professional theater of Broadway. They are the purveyors of art and science and culture for the millions upon millions—the promoters who represent the arts and sciences to humanity and who represent humanity to the arts and sciences; they are the means by which we are reminded of the famine in remote places or of war or trouble or change; they are the means by which this great earth and its peoples have become one to one another, the means by which the news of discovery or honor and the stories and songs of today travel and resound throughout the world. But they are also the means by which the true human community, the man knowing man, the neighbor understanding neighbor, the schoolboy learning a poem, the women dancing, the individual curiosity, the individual sense of beauty are being blown dry and issueless, the means by which the passivity of the disengaged spectator presents to the man of art and science the bleak face of unhumanity.

For the truth is that this is indeed, inevitably and increasingly, an open and, inevitably and increasingly, an eclectic world. We know too much for one man to know much, we live too variously to live as one. Our histories and traditions—the very means of interpreting life—are both bonds and barriers among us. Our knowledge separates as well as it unites; our orders disintegrate as well as bind; our art brings us together and sets us apart. The artist's loneliness, the scholar despairing because no one will any longer trouble to learn what he can teach, the narrowness of the scientist—these are unnatural insignia in this great time of change.

For what is asked of us is not easy. The openness of this world derives its character from the irreversibility of learning; what is once learned is part of human life. We cannot close our minds to discovery; we cannot stop our ears so that the voices of far-off and strange people can no longer reach them. The great cultures of the East cannot be walled off from ours by impassable seas and defects of understanding based on ignorance and unfamiliarity. Neither our integrity as men of learning nor our humanity allows that. In this open world, what is there, any man may try to learn.

This is no new problem. There has always been more to know than one man could know; there have always been modes of feeling that could not move the same heart; there have always been deeply held beliefs that could not be composed into a synthetic union. Yet never before today have the diversity, the complexity, the richness so clearly defied hierarchical order and simplification; never before have we had to understand the complementary, mutually not compatible ways of life and recognize choice between them as the only course of freedom. Never before today has the integrity of the intimate, the detailed, the true art, the integrity of craftsmanship and the preservation of the familiar, of the humorous and the beautiful stood in more massive contrast to the vastness of life, the greatness of the globe, the otherness of people, the otherness of ways, and the all-encompassing dark.

This is a world in which each of us, knowing his limitations, knowing the evils of superficiality and the terrors of fatigue, will have to cling to what is close to him, to what he knows, to what he can do, to his friends and his tradition and his love, lest he be dissolved in a universal confusion and know nothing and love nothing. It is at the same time a world in which none of us can find hieratic prescription or general sanction for any ignorance, any insensitivity, any indifference. When a friend tells us of a new discovery we may not understand, we may not be able to listen without jeopardizing the work that is ours and closer to us; but we cannot find in a book or canon—and we should not seek—grounds for hallowing our ignorance. If a man tells us that he sees differently than we, or that he finds beautiful what we find ugly, we may have to leave the room, from fatigue or trouble; but that is our weakness and our default. If we must live with a perpetual sense that the world and the men in it are greater than we and too much for us, let it be the measure of our virtue that we know this and seek no comfort. Above all, let us not proclaim that the limits of our powers correspond to some special wisdom in our choice of life, of learning, or of beauty.

This balance, this perpetual, precarious, impossible balance between the infinitely open and the intimate, this time—our twentieth century—has been long in coming; but it has come. It is, I think, for us and our children, our only way.

This is for all men. For the artist and for the scientist there is a special problem and a special hope, for in their extraordinarily different ways, in their lives that have increasingly divergent character, there is still a sensed bond, a sensed analogy. Both the man of science and the man of art live always at the edge of mystery, surrounded by it; both always, as the measure of their creation, have had to do with the harmonization of what is new with what is familiar, with the balance between novelty and synthesis, with the struggle to make partial order in total chaos. They can, in their work and in their lives, help themselves, help one another, and help all men. They can make the paths that connect the villages of arts and sciences with each other and with the world at large the multiple, varied, precious bonds of a true and world-wide community.

This cannot be an easy life. We shall have a rugged time of it to keep our minds open and to keep them deep, to keep our sense of beauty and our ability to make it, and our occasional ability to see it in places remote and strange and unfamiliar; we shall have a rugged time of it, all of us, in keeping these gardens in our villages, in keeping open the manifold, intricate, casual paths, to keep these flourishing in a great, open, windy world; but this, as I see it, is the condition of man; and in this condition we can help, because we can love, one another.

In the late fifties. (OMC)

7

AN INWARD LOOK

1958

I

THE CONFLICT WITH COMMUNIST POWER from time to time throws a harsh light on our own society. As this conflict continues, and its obduracy, scope, and deadlines become increasingly manifest, we begin to see traits in American society of which we were barely aware, and which in this context appear as grievous disabilities. Perhaps the first thus to come to attention is our inability to give an account of our national purposes, intentions, and hopes that is at once honest and inspiring. It is a long time since anyone has spoken, on behalf of this country, of our future or the world's future in a way that suggested complete integrity, some freshness of spirit and a touch of the plausible.

Two other national traits have more recently aroused grave concern. Because the conflict with Communist power is taking place concurrently with an extreme acceleration of a technological revolution, and in particular because these last years have marked the maturing of the military phases of the atomic age, public attention has been drawn to the relative effectiveness of the Soviet system and ours in the training and recruiting of scientists and technical people. This comparison has shown that, in a field where once we were better than the Russians, we may soon be less good. The Soviet system, by combining formidable and rare incentives for success in science and technology with a massive search for talent and with rigorous and high standards in early education, appears about to attract to scientific work a larger fraction of its population than we shall be doing.

When we learned this, it was natural to turn our attention to its causes. Some of these lie in the relatively low esteem in which learning is held in this country and, above all, in our indifference to the profession of teaching, especially teaching in the schools, a low esteem that is both manifested and caused by the fact that we pay our teachers poorly and our scientists not too well. The grimness of life in Soviet countries makes it easy to translate prestige into luxury and privilege. We do not want it so here. Yet on closer examination we have seen that in our own schools educational standards are far lower for languages, mathematics and the sciences than in their Soviet counterparts. We have learned that many of our teachers are not really versed in the subjects which it is their duty to teach and, in many cases, their lack of knowledge is matched by their lack of affection or interest. In brief, we have come upon a problem of the greatest gravity for the life of our people by matching ourselves against a remote and unloved antagonist.

Something of the same kind appears to be happening in a quite different area. This has to do with the ability of our Government—in fact, with the ability of our institutions and our people through our Government—to determine national policy in those areas that have to do with foreign affairs and strategy, military and political. To quote Mr. W. W. Rostow in an address to the Naval War College late in 1956:

> I do not believe we as a nation have yet created a military policy and a civil foreign policy designed to fulfill [our purposes] and to exploit the potentials for social and political change favorable to our interest within the Communist Bloc . . . Historically, the United States has thrown its energies into the solution of military and foreign policy problems only when it faced concrete, self-evident dangers.

Or again, Mr. Henry Kissinger wrote in the April 1957 issue of *Foreign Affairs*:

> By establishing a pattern of response *in advance* of crisis situations, strategic doctrine permits a Power to act purposefully in the face of challenges. In its absence a

Power will constantly be surprised by events. An adequate
strategic doctrine is therefore the basic requirement of
American security.

It is now a widely held view that, despite the organization of the
executive branch of the Government to cope precisely with long-
range problems, foreign policy and military strategy; despite the
role assigned to the Joint Chiefs of Staff, the National Security
Council and the Policy Planning Staff of the Department of State;
despite the availability to these organizations of the technical and
intellectual talent of the whole of this country and, to a more
limited extent, of the whole free world—despite all this, the
United States has not developed an understanding of its purposes,
its interests, its alternatives and plans for the future in any way
adequate to the gravity of the problems that the country faces.
There is a widespread impression that we live from astonishment
to surprise, and from surprise to astonishment, never adequately
forewarned or forearmed, and more often than not choosing
between evils, when forethought and foreaction might have
provided happier alternatives. Why should this state of affairs exist
in a country rich with wealth and leisure, dedicated to education,
with a larger part of its citizenry involved in education than in any
other land at any other time, with more colleges, universities,
institutes and centers than anyone cares to count, and at a time
when unparalleled powers in the hands of a dedicated and hostile
state threaten us more grievously than ever since the early days of
the Republic?

There are, of course, other national traits of which we can
scarcely be proud, on which neither the atomic age nor the
conflict with Communism has put much emphasis. We may think,
for instance, of our great wantonness with our country's resources;
we may think of the scarcity of instances in which a concern for
public beauty and harmony has made of the physical environment in which we live that comfort to the spirit which the
loveliness of our land and our great wealth could well make
possible.

Indeed, all of the traits in which we judge ourselves harshly
could have been drawn by historians comparing us with past
cultures, or observers of the current scene comparing us with
those contemporary. We should then, perhaps, have noted that no

91

people has ever solved the educational problem which we have put to ourselves, and that no government, in a world in which few governments succeed for very long, has ever succeeded in a problem of the scope and toughness of that which faces ours. Indeed we could recognize the traits of weakness in our society in terms of a norm or an ideal, and hear of them from the philosopher or prophet. I believe, in fact, that these ways are the more constructive, because I believe, as will be more evident in what follows, that the traits that bother us are signs of a rather deep, refractory and quite unprecedented cultural crisis, and that in the end they will yield, not to symptomatic therapy, but to changes in our life, changes in what we believe, what we do and what we value.

For the problems of our country and our age have hardly in historical times arisen in anything like their present form; certainly they have never been resolved. If our adversary appears to have solved them better than we, it may be healthy for us to note that; it can hardly be healthy for us to adopt his means. He knows what he wants, because he has a simple theory of the meaning of human life and of his place in it. With the strength of that confidence, he has a government prepared to take, at vast human cost, all necessary steps to reach his ends. That there is only a small, fragmentary, largely obsolete taint of truth to his theory, that it excludes the greater part of truth, and the deeper, should give us some confidence that he will not succeed. That his failure may be marked by a vast if not universal human involvement, and an unparalleled devastation and horror, should temper our pleasure in this prospect and return us to the solution of our problems on our own terms, in our own way, in our own good time.

For the traits of weakness in our society we can see grounds that are at once multiple, intelligible, and ironic. I think that the three weaknesses—in our education, in our faltering view of the future, and in our difficulties in the formulation of policy—have some common grounds; but they are not the same, and to follow them all is not the purpose of this paper. Certainly egalitarianism and our traditionally cherished tolerance of diversity, diversity precisely on the most fundamental issues of man's nature and destiny, his salvation and faith, certainly these qualities, long held as virtues, have much to do with our troubles in education where they define, as it were, the insoluble problem; they have much to do with the difficulties of prophecy and policy, which traditionally rest on consensus precisely with regard to those matters where we are dedicated to difference. The good fortune of the country, speaking in large terms and over the centuries, and its consequent optimism and confidence, have something to do with our troubles. Perhaps we would not change these things, but we must give weight to them, when we compare ourselves with Athens, or Elizabethan England, or Victorian, or seventeenth-century France.

Our weaknesses, of course, have a touch of irony. It is our very confidence in education, our determination that it should be available to all, our belief that through it man will find dignity and freedom, that have played so large a part in reducing our educational system to the half-empty mockery that it now is. When, for the first time in years of formal peace, we have devoted effort, study, thought, and treasure to the quest for military security, we have brought about the most fearful insecurity that has been known to man in what we know of his history.

II

It is commonly said that our national culture favors practice over theory, action over thought, invention over contemplation. There is some truth to this thesis. It should not be exaggerated. For one thing, the balance between operation and reflection must always, everywhere, numerically favor the doers as compared to the reflectors; even in Athens there were quite a few Sophists for one Socrates; and I find it hard to imagine any society in which the world's work does not occupy more people more of the time than does an understanding of the world. For another, the balance between these aspects of life has been accented by circumstance, in that the doers in our country have had great good fortune to mark and celebrate their deeds; the country's wealth, its spaciousness, its wide measure of freedom, and, on the whole, its prevailing optimism. It would take quite considerable accomplishments of theory and understanding to match the brilliance, often almost the impudence, of our material creations.

Our past has always been marked by a few original and deeply

93

reflective minds whose work, though it was part of the intellectual tradition of Europe and the world, has nevertheless a peculiarly national stamp, as in the four names of Peirce, Gibbs, James, Veblen. Today, in almost all fields of natural science, and in some others as well, our country is preeminent in theory as it is in experiment, invention and practice. This has meant a great change in the educational scene, as far as higher education is concerned, in the graduate schools, in post-doctoral work, in the institutes and universities. Part of this, it is true, has come about because of misfortunes abroad: the two wars in Europe, and the Nazis, the initial effects of Communist power in Russia, which for a time at least made conditions of serious study very difficult. It has been brought about in part by the coming to this country of scholars in refuge from their regimes, from tyranny and trouble abroad. Nevertheless it is true that today a young man wishing the best training in theoretical physics or mathematics, theoretical chemistry or biology, will be likely to come to this country, as three decades ago he would have gone to the schools of Europe. It was important, after the end of the Second World War, when there was much public interest in the successes in applied science which the war years had brought about in this country, to combat any exaggerated sense of American superiority by pointing to the great contributions for which we were in debt to Europeans and others from other lands; but to repeat today that which was only partially true then, namely that Americans excel in practical undertakings but are weak in theory, is to distort the truth. It should be added, of course, that the number of men engaged in theoretical science is always small and, even with us today, it is very small. Their work and their existence can have little direct bearing on the temper and style of the country.

Having said all this, it does seem to me that in comparison with other civilizations—that of classic India surely, that on the continent of Europe, and probably even that of England, where theory is brilliantly made but largely ignored in practice—ours is a land in which practice is emphasized far more than theory, and action far more than contemplation. In the difficult balance of teaching, we tend to teach too much in terms of utility and too little in terms of beauty. And if and when we "do it ourselves," it is unlikely to be learning and thought.

To see the bearing of this trait, we should recognize another feature of the American landscape: in important, deep and complex ways, this is a land of diversity; and it tolerates, respects and fosters diversity in the form of a true pluralism. There is much theory made in the United States: cosmological theory, theory of genetic processes, theory about the nature of immunity, theory about the nature of matter, theory about learning, about prices, about communication; but there is no unifying theory of what human life is about; there is no consensus either as to the nature of reality or of the part we are to play in it; there is no theory of the good life and not much theory of the role of government in promoting it. The diverse talents, skills, beliefs, and experience of our people contribute effectively to the solution of a concrete problem, to answering the well-defined question, to the building of a machine, or a structure, or a weapon system; and in such concrete and limited exercises, the diversity and strangeness of the participants is harmonized by the community of the concrete undertaking. The team of experts, sometimes including experts from social science, was an immensely successful invention for wartime research, and continues to be in many forms of technical enterprise. It continues to be inappropriate, and tends to languish, in the general undertakings of academic life.

American pluralism can no doubt in part be understood in terms of our history, and those features in which we differ from most of the communities of Europe and of much of Asia. We may think of the relatively primitive communities in the Indian villages of the Southwest, which some of us may still remember from the earlier years of this century. The quality of their life was relatively static and highly patterned; all of its elements were coherent, and were rendered unified and meaningful by religious rites and religious doctrine. Change was slow, and communication adequate to the limited experience of the villages. Such communities represent almost an ideal of unity, of common understanding, and of a monistic view of the world. There has been little of the village in American life. The frontier, the openness of the country, and later the immense rapidity of change and the tumult of motion and traffic, have given us a very different national experience. Probably for two centuries New England had the stability of village life; and I believe that we see today, in the coherence, firmness and mutual understanding of its survivors, one of the most stable and unified elements in our country.

95

Probably, although I know less of this, one could find a similar story in the South, though the fortunes of the last hundred years have dealt harshly with it.

Even if we turn our thoughts to Europe, the site of so much of the commotion, disillusion, and variety which characterize our own land, we see important differences; there is a long past of limited mobility, culminating in the thirteenth century in the unified view of all matters important to man, in a universe determined by God, with God omnipresent, with the unvarying natures of all finite things, and the ever-present end and purpose of man's life. When this world began to break, it broke slowly, first in the minds of the philosophers and scientists. It was not until the seventeenth century that the turn from contemplation to action can be seen with any completeness; long after it occurred, its consequences were still troubling to John Donne: "'Tis all in peeces, all cohaerence gone. All just supply, and all Relation." Man's awareness of his power came slowly to Europe; it came to people bound by a common tongue, a common habit, and common traditions in taste, manners, arts, and ways.

Compared to all this, Americans are nomads. There is, of course, much in common in what brought people to this country; but in overwhelming measure, what was common was either negative or personal and practical: the desire to escape repression, or the hope of making a new fortune. In the formative years of our history, emptiness, the need and reward for improvisation, variety, and the open frontier endowed the differences between men with weight and sanction. Our political philosophy undertook to reconcile the practical benefits of union with the maximum tolerance of diversity. To all of this has come within the last century, and complementing the closing of the physical frontier, a new source of change, more radical and in the end more universal than those before. This lies, on the one hand, in the unprecedented growth of knowledge, whose time scale, estimated apprehensively as a half century two hundred years ago, could better now be put at a decade; and with this, based partly upon it, partly upon accumulated wealth, and partly on the tradition of freedom and mobility itself, a technological explosion and an economy unlike any the world has seen.

Early in this century, William James wrote:

The point I now urge you to observe particularly is the part played by the older truths . . . Their influence is absolutely controlling. Loyalty to them is the first principle—in most cases it is the only principle; for by far the most usual way of handling phenomena so novel that they would make for a serious rearrangement of our preconception is to ignore them altogether, or to abuse those who bear witness for them.

In our time the balance between the old truths and the new has been unhinged, and it is not unnatural that most men limit, in the severest possible way, the number and the kind of new truths with which they will have to deal. This is what makes the intellectual scene a scene of specialists, and this is what makes our people, for all the superficial evidences of similarity, more varied in their experience, more foreign to each other in the tongues which they use to talk of what is close to them, than in any time or place which comes to mind; this is what limits consensus to statements so vague that they may mean almost anything, or to situations so stark and threatening and so immediate that no theoretical structure, no world view, need intervene.

Perhaps the most nearly coherent of all our large theoretical structures is that of natural science. It is hardly relevant to many of the questions of policy and strategy with which our Government must be confronted; to some it is. This coherence is, however, of a very special sort: it consists by and large in an absence of contradiction between any part and any other, and in a pervasive, often only potential mutual relevance. It does not consist in a structural coherence by which the whole can be derived from some simple summary, some key, some happy mnemonic device. There are thus no fundamentals of science. Its largest truths are not definable in terms of common experience; nor do they imply the rest. Our knowledge of nature is in no true sense common knowledge; it is the treasure of the many flourishing specialized communities, often cut off from one another in their rapid growth. Never has our common knowledge been so frail a part of what is known. Natural science is not known, and probably cannot be known, by anyone; small parts of it are; and in the world of learning there is mediation in the great dark of ignorance between the areas of light.

In assessing the practical import of scientific developments, the Government may be faced by a reflection of this situation. Even in so relatively limited a field as the peacetime hazards of atomic radiation, it cannot turn to an expert for the answer. It turns to the National Academy of Sciences, which assembles a series of committees, both numerous and populous, whose collective knowledge and collective recognition of ignorance is, for the time being, our best answer.

In other aspects of intellectual life, more relevant to policy and to strategy, we find a situation not wholly dissimilar, though less formalized and less clearly recognized. In our own internal affairs, knowledge on the part of the Government of what the situations in fact are with which it must deal is complemented by a traditional safeguard in our political institutions. If, in fact, the executive and legislative branches of the Government have erred in their assessment of the problems of Northwestern lumbermen, or of maritime labor, or of Marine recruits, there is opportunity for those who are specialists in these ways, because they live in them, to be heard; and there is an underlying tolerance, sometimes violated, sometimes ignored, and most intimately and immediately knowledgeable, the grave weight of the doctrine of the concurrent majority. In foreign affairs, in matters affecting other lands and people, no such protection and no such redress exist. Here the Government must rely most heavily on what is essentially scholarship: what the historian, the linguist, the artist and all others who, with the slowly learned historian's art of judging, evaluating and understanding, can give as an intimate glimpse of what goes on in foreign and often very strange lands.

Faced with all this, faced with the complexity, the variety, and the rapid change which characterize both the intellectual scene and the world itself, there is a terrible temptation to seek for the key that is not there, the simple summary from which all else might follow. We have tended to do that in the wars of this century, with, it would seem most probable, consequences of great trouble when we have come to the end of the war. It was probably bad even in the First World War, when our Government had a relatively elaborate and learned theory which was widely accepted by our people, but which was not quite true. It was probably bad in the Second World War, where the theory seemed to be very primitive and to consist of the view that evil, however widely spread in the world, was so uniquely concentrated in the governments of the hostile Powers that we could forget it elsewhere.

A government may, for more or less valid reasons, reach a conclusion as to what its action should be, as ours does when we declare war, or when we adopt such relatively well-defined policies as the Truman Doctrine. Such decisions, reflecting the best estimate of the evidence available when they are made, are acts of will; clearly, further evidence which supports the decisions reinforces the will, makes the prosecution of the war or the execution of the doctrine more likely to be effective. Evidence that the decisions may have been in error or may no longer be timely has a contrary effect. The human commitment to its own decisions, the human reluctance to learn and to change should not be reinforced by any doctrine which deprecates the truth, and therefore the value, of what is inconsistent with past evidence and past judgment. The danger lies, not so much in that the new and conflicting evidence may be weighed and given too little weight; it is that it will not even be seen, that our organs of intelligence and perception will be coded, much as our sense organs are, by our commitment, so that we will not even be aware of inconsistency and novelty.

I believe that we are now deeply injured by the simplifications of this time. The cold war is real, it is bitter, and it is deadly. But it is not the only issue in the world, and for countless other peoples and their governments it is not the issue they see in the brightest, harshest light. Such global views tend to inhibit the reception of essential knowledge because in the light of our dominant doctrine this knowledge appears irrelevant or somehow does not fit. That we are indeed in this danger seems to me clear from the extent to which the unfolding of history finds us always surprised.

There are two features of the situation that I have attempted to sketch that need a special comment. It seems to me that both the variety and the rate of change in our lives are likely to increase, that our knowledge will keep on growing, perhaps at a faster and faster rate, and that change itself will tend to be accelerated. In describing this world, there will probably be no synopses to spare us the effort of detailed learning. I do not think it likely that we are in a brief interval of change and apparent disorder which will soon be ended. The cognitive problem seems to me unprecedented in scope, one not put in this vast form to any earlier society, and one for which only the most general rules of behavior can be found in the past.

It also seems to me that we must look forward to a world in which this American problem is more nearly everyone's problem. The beginnings of this are perhaps as important in the present moods of Europe as are the history of the two Great Wars, Communism, the Nazis, and Europe's loss of political, military and economic power. The problems seem clearly implied in the determination of peoples in Africa and Asia, and in Central and South America, by means not yet devised and not at all understood, to achieve education, learning, technology, and a new wealth. They form a part of the unrest, newly apparent in the intellectuals of the Soviet world, perhaps especially among their scientists, and increase the somberness of any prospect of change from tyranny to freedom.

There are thus the most compelling external reasons why we, in this country, should be better able to take thought, and to make available in the pressing problems of policy and strategy the intellectual resources now so sorely lacking. They are needed in the struggle with Communism; they are needed if we are to have some understanding and some slight influence, in all the rest of the world, in the great changes that lie ahead for it. Awareness of this need will do us good; and I do not underestimate the value of its general recognition by the people of this country nor official recognition by their Government. It can only help to make money available to education and to teaching; it can only help to make the learned as well as the facile welcome in the proceedings of government policymaking. But though these measures are bitterly necessary, and though they are long overdue, the real thing will not, I fear, come from them alone.

There may be valid grounds for a difference of opinion as to whether an official recognition of a need, or even a generally understood recognition of a need among our people, will evoke the response to that need. What we here need is a vastly greater intellectual vigor and discipline; a more habitual and widespread openmindedness; and a kind of indefatigability, which is not inconsistent with fatigue but is inconsistent with surrender. It is not that our land is poor in curiosity, in true learning, in the habit of smelling out one's own self-delusion, in the dedication and search for order and law among novelty, variety, and contingency. There is respect for learning and for expertness, and a proper recognition of the role of ignorance, and of our limits, both as men and as man; but of none of these is there enough, either among us, or in the value with which they are held by us, if indeed government by the people is not to perish.

Robert Oppenheimer with his mother, around 1906. (OMC)

8

TRADITION AND DISCOVERY

1960

[...] WHEN COLUMBUS SAILED on his first voyage, his first voyage of discovery, it is told that the first evening with the ship standing out to sea he opened the pages of what would later be the log of this voyage, and on it he wrote *Jesus cum Maria sit nobis in via*.

This sense that we are entering a new time, that whatever we owe to the past, and however all our future is built on it, that we have problems unlike those or somewhat unlike those that have ever been dealt with, is very deep in my mind. Terror attaches to new knowledge. It has an unmooring quality; it finds men unprepared to deal with it. Think of the story of Adam, and think of the story of Prometheus. They are very different but they have this in common. Even in discoveries which affect our ideas and which do not immediately or visibly affect the state of men's lives, even in abstract discoveries, there is this same sense of terror. I have found it among my colleagues and have recognized it in myself.

This question of what discovery does to tradition and how tradition makes discovery possible is, of course, a special case of a problem that is very old: the problem of the struggle and the conflict and the balance between what is familiar and essentially timeless in our lives, and the always manifest and now overwhelming sense of change. For tradition, its whole effort is to preserve, to refresh, to transmit, and to increase our insight into what men have done as men, their art, their learning, their poetry, their politics, their science, their philosophy. Tradition is no less than what makes it possible for us to deal as sentient and thinking beings with our experience, to cope somehow with our sorrows, to limit and ennoble somehow our joys, to understand what happens to us, to talk to one another, to relate things to one another, to find the themes which organize experience and give it meaning, to see the relevances of things to one another. It is, of course, what makes us human and what makes us civil. It is typically and decisively the common heritage, that which men do not have to explain to each other, that which in happier days they did explain to their children, that which they can rely on as being present, each in the other's head and heart. It points to the connection of things. Tradition, of course, is always a very oversimplifying thing, since things, in fact, are not completely alike, and not completely related. It finds the great human themes which run through everything, which we can come back to, which we can recognize, which we can communicate. This communication, often verbal, but of course, by no means necessarily so, is the heart of a human community, and the essence of human life.

[. . .] Tradition is also the matrix which makes discovery possible. It is the organ of interpretation, of enrichment and understanding, that in the arts, in the sciences, and even in our common ethical life gives meaning to new discovery. It provides the tools of new discovery. It is, of course, the special mark of modern European tradition that it has catalyzed, and no one fully understands why, an immense outpouring and an immense growth of discovery unlike anything which man has known. It is an unprecedented use of the past for the future, an unprecedented enrichment of the power to find new things, by virtue of the extent to which we are in control and have some understanding of the old, unprecedented in volume, in weight, in wealth, and in scope, unprecedented in many ways even in quality, even if one thinks of the highest days of ancient times.

I regard our situation as grave, interesting, and radically novel, something which people have not had to face in man's history. It will put difficult choices to us; it is doing so now. And we can be judged, we will be judged by our response. We can be judged both in our own countries, in what we make of them, and by the sort of example that we set for the larger world.

Science rests on, intersects with, alters, affects almost all of man's ethical life. The change in the world which its growth has made, both material and intellectual, is an unfathomably great one. I do not propose to address the material changes. I have lived through many in my life and those who are young will live through many more. They are familiar; they are important. On the intellectual side, however, there are three traits which should be discussed. One is the growth of knowledge itself; one is the question of its structure; and one is a related question, the openness of knowledge—its potential infiniteness and the fact that it confronts us with choice. [. . .]

All mature men today are really necessarily and deeply quite unaware of the greater part of what is known. We did not learn about it in school; we have no immediate practice in it. It involves a way of talking, and a tradition, for which we are barely prepared. These have grown out of what was learned, when we last looked at the subject. Sometimes we suddenly are shocked by a recognition of how fast things move. Someone tells us that as far as an understanding of life is concerned, we have learned more in the last five years than in all the history of man. Sometimes some practical thing like atomic energy shocks the public generally at how much was going on, not just in secret but in private; going on and nobody knew about it, until it changed the face of the world, maybe for the worse, maybe for the better. Of course, we live in a time when this problem is compounded—and we hope that this will be true quite widely—by an egalitarian and open attitude toward the acquisition of knowledge, where accidents of birth and fortune—which to some extent will be reduced, and very largely have been reduced compared to a hundred years ago—will play very little part in deciding how much a man learns, how much he studies, what right he has to enter the life of the mind, and to spend his life there.

[Scientific] knowledge is not just a collection of miscellaneous, unrelated things. It is not without order; order is what it is about. Its purpose is to discover, and in order to discover, to create the order which relates things with one another, and to reduce—though it certainly never eliminates—the arbitrary in our experience. It is not orderly in the sense that there are a few general premises from which everything else follows. One cannot say: "But of course, I really don't know about nature and man, but I know all the basic principles; the rest I could pick up." The deep things in my science, the deep things in mathematics and increasingly the deep things in biology, are not things that one

105

knows about. They are very hard to learn, and they are built on a long, long, cumulative structure of a specialized tradition. There is a lot of relation in this world of science, and it has structure and refers to a world full of order, and it is rich and astonishing and subtle. There is order so that things cohere, and general things encompass special ones; this means, in fact, that a great deal of what was in textbooks a long time ago does not have to be in them today. It means that there is a kind of sloughing off, not because things stop being true, but because one can remember some general rules from which particular things follow. [. . .]

There is another sense to this kind of unity, and it is important that we be aware of it. No part of science follows, really, from any other in any usable form. I suppose nothing in chemistry or in biology is in any kind of contradiction with the laws of physics, but they are not branches of physics. One is dealing with a wholly different order of nature. What is simple in one, is complicated in the other. I think that probably over the last decades the great synapses which separate one branch of learning from another have tended to yield. One sees how the connections can be made, for instance, about how life could have originated from inert matter, and how it transmits itself. Thus the whole idea that a necessary cause, an efficient cause, could be consistent with purpose is illuminated with the idea of purpose. It is illuminated by this, so that the characteristic features of life—which are that life has to be described in terms of ends and purposes—are not in conflict with the ideas of necessity, the ideas of efficient causation, and the universal validity of the large and lovely laws of physics.

The receptacle of all this knowledge is, of course, not man in general, nor is it quite the individual specialist. Rather, it is the specialized communities of interlocking expertise, men who have very limited titles so that it may take a lot of words to say what they do, like high energy physicists or high polymer physicists. To these groups the knowledge comes. They have close professional relations; they typically rejoice in any success that anyone else has had, even though they may wish that they had had the success themselves. This is a hallmark of science. If people are jealous of each other more than they rejoice in what the other finds, then somehow this has nothing to do with science.

There is an intimacy in these communities, and it always has given me a very high standard for the intimacy of men with each other, and a forlorn hope that this sort of intimacy could help to hold the world together while our political institutions slowly adapt to the changed circumstances in the world. We have a sort of modern version of the medieval guilds, a kind of syndicalism, of cognitive syndicalism.

[. . .] Knowledge is twinned with ignorance. Knowledge excludes knowledge. When you know something or learn something, you in doing that, put aside, lose, perhaps not permanently but in the context, the whole opportunity of finding many other things. Instead of making a long abstract talk about this, I shall give four examples that seem to me to illustrate this. [. . .]

The first is an experiment which was done by Jean Rostand in Paris. It has to do with the signals that are tapped off the auditory nerve, let us say, of a dog. He did it with many animals, and he made a record of these signals. Every time he rang a bell he got a recognizable signal. If he put a piece of meat in front of the dog, and then rang the bell, there was no signal at all. The higher centers instructed the sense organs themselves not to record this event of the ringing of the bell. This is instructive when one reflects on how people have tried to build their sense of objectivity on the sense datum and the immediacy and reliability of what one learns from one's senses.

A second example is a little different. This was done by Hebb and his associates at McGill. He was actually preparing to make a study of brainwashing during and after the Korean War. He wanted to see what happened to people when you did nothing to them. His idea of nothing was this: he would put a man in a quiet room, a soundproofed room, in which only a low hum was kept on all the time. The man's eyes were covered so he could not see anything. His fingers were covered with cotton batting, and a cuff to protect his sense of touch. Hebb asked for volunteers among medical students and other presumably normal people; and they all thought that this was wonderful, that they would get clear all the points that they had been puzzled about, and be left alone. He told them that he would keep them alive with adequate food, and he had no trouble getting volunteers. They stayed in from something like a day or day and a half to a maximum of six days. When they came out they had lost their rational faculties—not permanently, but for a time which was somehow related to the length of time they had stayed in this condition. They could not add, they could not subtract, and they could not see an organized perspective in which we have sense of where things are, they had other troubles with words and so on. Apparently the availability of the simplest rational faculties depends on maintaining some kind of constant traffic.

The third example is even more revealing. It was carried out by a man called Land, near Boston, also for practical purposes, because he is the head of the Polaroid Corporation and he wanted to get a film which could be quickly developed and show color; thus he studied color vision. The theory we have had in the past is that there are three primary colors and that there are pigments in the retina which respond to these colors, and that when the corresponding wavelengths of light hit the retina we recognize red, or blue, or some combination of them which may be purple. This may be true; but Land showed that it is a very small part of the story. If you take a multicolored scene and you illuminate it with yellow light, say with the sodium line, which is used so much in laboratories, and make a black and white print of it, and then if you take another yellow line only a few percent different in wavelength, a couple of hundred Angstroms different, and take a picture of the same multicolored scene and make a black and white print of it, and project these with the light that you used to take them, then people like you and me see color. We see red, green, blue, yellow, everything that was in the scene, although none of that light is coming to the eye. As a matter of fact, we see it if we put one image on one eye and one image on the other. This understanding of the meaning of these two signals in two almost identical colors of yellow, is performed, not in the eye, but some place further up toward the cortex.

The last example I cannot explain to you. It has to do with the fact that one of the traits which we are familiar with in the simplest physics of large-scale bodies suddenly went away. Since before the time of Newton, but with much more confidence since the time of Newton, we talk about the motion of bodies by giving at some time their location and their velocity, and then, if there are no forces acting, the conservation of impetus, of momentum says that the velocity will not change. If there are forces acting, then an orbit will be described like that of a baseball or projectile on earth or of the planets circling the sun. Newton managed to unify in his laws the enormous experience going from celestial bodies to quite small terrestrial experience. One cannot go on indefinitely, because if one looks at the atoms themselves, at the electrons which give them all the properties that we normally notice—their color, their chemical properties, their ability to form compounds, their ability to form solids, and crystals, and so on—then it is no longer true that one can get any kind of agreement with the observations of nature, if one makes the assumption that even if one does not know it, the electron at some time has some position and some velocity, and that if you know one, you may suppose the other unknown to you but existent. If one assumes that, one gets in gross contradiction with the existence of stationary states, and with all the properties of atoms that I have enumerated. The great discovery was that you cannot attribute such properties as position and velocity and energy to such a system, unless you have taken the trouble to make an experiment in which you are prepared to measure what this quantity is. It is not that you do not know it; it is that you cannot assume that it exists. The attempt to objectify this, and say, it is there, but I am not clear just what it is, leads to disaster. This, which is the heart of quantum theory, the theory of complementarity, still further reinforces the sense of option in knowledge, the sense that you have a choice as to which study you make; the sense that having made that option, there is a kind of indivisible whole to the affair. You cannot go back on your bargain without spoiling everything. It illustrates further that objectivity consists not in the fact that it is there independent of our experiment with it; it is there precisely because we can tell each other what experiment we did and how it came out.

I have had extremely bad luck with this last point in trying to make it clear either in the course of a lecture or of many lectures. One reason is that it does take a pretty good knowledge of physics to know what the words mean and not to misinterpret them—indeterminacy, for instance, comes up in this connection, and relativity; and they sound very much like a state, a mood that we are in most of the time. They do not really mean that. One gets the desperate feeling that any words that are familiar will become puns, and any words that are unfamiliar will take a very long time to explain.

[. . .] We can hardly expect that a science as old and mature as physics has a very direct, immediate bearing on philosophical thought, or on the common discourse on which philosophical thought is based. It seems to me that it needs first to be far more intelligible, as for instance, in elementary terms, the three first experiments I mentioned seem to me intelligible by ordinary people, not highly trained specialists. [. . .]

There is also a second requirement. It is not enough that something be intelligible for it to enter in a major way into the philosophy of the times, as one says of Newton's mechanics that it did enter the eighteenth-century idea of man, his place, and even his dealings with his fellows; and, as an even better example, in Darwin a hundred years ago. It is also necessary that there be a kind of sensed relevance of what has been discovered in science, to the aspirations, the interests, the direction and the hope of the society. There has to be an entering into the discourse of cultivated men—and this I hope will mean, more and more, almost everybody—a digestion and transmutation and adaptation of the new lore about nature or about man. It is likely that it will also turn up as something new in philosophical speculation, but that, in turn, is historically and typically based on the common talk of the time. It can go beyond it, but it cannot be unrooted from it.

[. . .] What, then is the relation between the scientific explosions of this age and the weight and the excellence that we may all hope to achieve in common discourse? I have in mind an image of common discourse which is itself blurred by three related realities. One is the size of our world and its great communities, the number of people. One is the generally egalitarian and inclusive view that there be no *a priori* restrictions on who is to take part in the discourse. Not everyone will, but I think our civilizations, with all their differences, share the hope that everyone may. And the third is the extraordinary rapidity with which the preoccupations and the circumstances of our life are altered [. . .].

I would like to make a few comments on the nature of the relations between rational discourse, culminating, in good times, in philosophical discourse on the one hand; and the development of science on the other. Of course, I think of the country I know best, and the experiences we have had, of how we use our leisure, and how we talk, and what we think. I apologize in a way for this provincialism. We are further along on the road of very high productivity and very high consumption than most countries, and we have gotten there without giving very much thought to the quality of public life while we were on our way. I am far from clear that some of the more acute troubles which we have at home may not be softened or even averted. I see troubles like this coming in Europe, and I think the best we can hope, the most we can hope, for the Communist world, is that it be free enough for such troubles to come there.

I need hardly bring to mind that the great sciences of today arose in a double source, philosophical discourse and speculation, and in technical invention. One could not have done without both. Different men will trust one or the other. Why has the enormous success, unanticipated, not fully appreciated, and never fully realizable, success of one sort of intellectual activity, science, not had a beneficial effect on the intellectual life of man? In some ways it has. It has put an end to many superstitions, much darkness. But if we think back to the early days, either of the European tradition or of modern society, we see very few people in this process, this discourse: the citizenry of Athens, the few handfuls of men who concerned themselves with the structures of American political power, the participants in the enlightenment in Europe in the eighteenth century, from Montesquieu to the Revolution. They are not very many men. They have before them a relatively well-digested and common language, experience, and tradition, and a common basis of knowledge.

If we look today, we see a very different situation, an alienation between the world of science and the world of public discourse, which has emasculated, impoverished, and intimidated the world of public discourse with only limited countervailing advantage. Thus any man may say what he thinks, but what he says is denied in public discourse any true element of legitimacy; it has a kind of arbitrary unfounded quality.

There was a relatively stable and a deeply shared tradition, an historic experience which was common among the participants in the conversation, and sometimes, even a recognition, of a deep difference in kind between the kind of use and value which public discourse has as its high ideal, and the kind of criteria by which the sciences themselves in part must judge themselves, must judge their truth. Of course, the sciences are rooted in public discourse. They embody taste, a sense of beauty, of simplicity, of order, and of depth. These are words you find always in any criticism or

111

appreciation of science. Indeed a lack of recognition of this is one of the reasons why the human quality of the great sciences has been so little understood.

[. . .] There are things important to discuss and analyze, to explore, to get in order in a certain sense, things, which are not best viewed as propositional truth, which are not assertions, verifiable by the characteristic methods of science, as to the existence in the world of this or that connection between one thing and another. They have rather a normative quality and a thematic quality. They assert the connectedness of things, their relatedness, their priority; without them, there would be not only no science, but no order in our lives.

We all know how great the gulf is between the intellectual world of the scientist and the intellectual world, perhaps hardly existing today, of public discourse on fundamental human problems. Some of the characteristics in the growth of science have closed it off from contributing to our talk with one another. Our usual ideas about objectivity may not be right, may not be relevant in this context. The fantastic growth, fantastic specialization, and the non-hierarchical character of science, means that one cannot easily master or compact it. This is a set of circumstances which has largely deprived our public discourse of its first requirement: a common basis of knowledge. But I believe that in excluding this kind of order, and this kind of verifiability, one has impoverished such public discourse. It is a very hard thing to say: "I leave out, I leave aside, I leave as irrelevant," something which is as large, as central, and as humane, and as moving a part of the human intellectual history as the development of the sciences them-selves. [. . .]

As to the question of a stable, shared tradition, I have, of course, been talking about philosophy in a culture like ours, which is predominantly secular, and have not talked of revelation as a living, present universal thing. It is not that I wish to exclude it; in a country which has an explicit separation of church and state, and an explicit freedom of religion, and a strong Protestant tradition, with its highly individual relation of man to his God, it is just not practical to talk in any other terms. Our tradition is buffeted by the most appalling and rapid changes. We all know how unprepared fifty years ago the world was for the tragedies of the twentieth century when it opened, and how bitter, corrosive, and indigestible many of them have been. I think primarily of the two World Wars and the totalitarian revolutions. We certainly live in the heritage of a Christian tradition. Many of us are believers; but none of us is immune from the injunctions, the hopes and the order of Christianity. I am not happy over the fact that one part of the Christian tradition seems almost absent. There has been no ethical discourse of any excellence or nobility of weight, dealing with how one should handle, how one should regard the new weapons of war, the atomic weapons. I welcome the fact that there is some discussion; but it tends to be discussion of a prudential kind: what will the enemy do, if we do that? I am glad that there is some talk, because five or ten years ago only those who were very close to the subject talked about it. But what are we to make of a civilization which has always thought of ethical questions as quite essential in human life, and which has always had a deep articulate, fervent conviction, probably never a majority conviction but always there, never absent, that returning of good for evil was the right way to behave, what are we to think of such a civilization which has not been able to talk about the prospect of killing almost everybody, or everybody, except in terms of calculation and prudence? That is not in any sense ethics. I would go only so far: in all those instances in which the West, and this is mostly the United States, has expressed the view that there was no harm in using the super weapons, using them massively, provided only that they were used against an antagonist whose government had done something wrong, we have made a very grave mistake. Our lack of scruple, which has its historical origins, not wholly, but largely, in the numbing and indifference which the terrible conduct of the Second World War brought to us, has been a great disservice to the cause of freedom and of free man.

[. . .] The attainment of certainty is not always the purpose of discourse. Its purpose may be the exploration of meaning. It is what we wish, what we intend, what we hope, what we are prepared to do, what we cherish, what we love, what we worship. My belief is that if the common discourse can be enriched by a more tolerant and humane welcome for the growth of science, its knowledge, its intellectual virtue, it may be more easily possible to accept the role of clarification and of commitment which is its true purpose.

I would think that we could look to a future in which very high on the list of purposes, of that leisure, of that consumption which our economies will gradually bring to us, was knowledge and thought; a future in which intellectual vigor of man had a greater scope than at any time in history, where man is free to love, to live, and to know. [. . .] One may hope that we will be unintimidated by the growth of knowledge and prepared for change, and ready to devote, not four years in college, not six years in graduate school, but our whole lives to an intellectual vigor and a true life of the mind. One may hope that we will learn again to talk with one another and share our sorrows and loves with one another and speak of meanings and the great themes of our lives and life. For if it is part of our problem today that to know and to discover one must act and choose, it is also another part of it, that, to act and to live, we must speak to one another, and we must hear.

At a panel discussion in 1961. (Photograph by P. Karas, courtesy of MIT Museum)

9

PROGRESS IN FREEDOM

1960

[. . .] TO ASSESS THE PROSPECTS of progress in liberty of the decade ahead, it is appropriate to look at the great changes of the decade just past. How great they are, how the very conditions of our lives have altered, reminds us of the central feature of our time: in the span of a man's life, we live many lives, in many worlds. A decade ago, for instance, Berlin and almost all of Europe, still bore everywhere signs of the ravages of war. Berlin, like much of Europe, has in some sense recovered. Its economy and prosperity could hardly have been anticipated a decade ago. But the greatest change in this city is that, in one respect, there has been no change. Its citizens live with a government and a style of life very largely of their own choice. [. . .] Ten years ago the Korean War, surely in the making, had not yet broken out; the guns that were to open that limited but most bitter conflict had not yet spoken. Ten years ago one could hardly have imagined that this spring and summer some dozen newly constituted nations would be on the point of seeking membership in the United Nations Organization, nor that the quest for national independence, for rapid modernization, and for appropriate regional or cultural international cooperation could have progressed so far and so fast. [. . .]

Ten years ago Stalin ruled Russia; in China the new Communist government was at the beginning of its consolidation of power. Ten years ago there was an almost total barrier to cultural and technical communication between the scholars of the Communist world and the West. Ten years ago it could still be argued what

117

vitality and what promise would lie in the gradual creation of a united Europe.

Among all the changes of this strange decade there are two which stand out. One is brutal. Ten years ago my country had barely lost, and still effectively had, a monopoly of the great new weapons, the atomic weapons. For their use in combat our armed forces, and all others, had means of delivery not essentially different from those of the Second World War. Yet it was then generally held, and I believe correctly, that these armaments constituted for all of us a hideous argument against the outbreak of general war. Today there can be no talk of monopoly: we are deeply into the atomic age, in which many nations will be so armed.

In this decade the deadliness, the destructive power of atomic stockpiles has increased far more than a hundredfold—how much more, it may be neither permissible nor relevant to tell. Today, the new means of delivery and use have made of the command and control of these weapon systems a nightmare fully known only to those responsible. They have added change to anger as another cause of disaster. What some of us know, and some of our governments have recognized, all people should know and every great government understand: if this next great war occurs, none of us can count on having enough living to bury our dead.

This situation, quite new in human history, has from time to time brought with it a certain grim and ironic community of interest, not only among friends, but between friends and enemies. This community has nothing to do with the injunction that we love our enemies, but is a political and human change not wholly without hopeful portent.

The Bhagavad Gita, the great Hindu scripture, is a sustained argument on the nature of human life and its meaning, introduced by Prince Arjuna's reluctance to engage in fratricidal combat. Vishnu describes this combat as a simple and necessary duty, whose performance would preserve the way of Arjuna's salvation, and whose evils were of no deep meaning, either for him or for those whom he might kill. Can we be thus comforted?

Traditionally, the national governments have accepted as their first and highest duty the defense and security of their peoples. In today's world they are not very good at it. We all know that the steps which we have taken, alone or in concert, have at very best an uncertain, contingent, changing, and above all transitory effectiveness. This is one reason, important but perhaps not central, for a second change in this past decade. We have come to doubt the adequacy of our institutions to the world we live in; beyond that, we have come to doubt certain aspects of the health of our own culture. In this, I speak with my own country in mind, because the traits that have given rise to our anxieties are as marked with us as anywhere. Yet I think I see that in the older, more traditional societies of Europe, the same problems are beginning to appear, and will inevitably grow more grave. I think that I see that in the measure in which productivity, education, and the modern world come to the peoples that aspire for them, these problems in their own form, will come too.

Compared to any high culture of the past, ours is an enormous society. It is for us an egalitarian one, in which we hope—and I pray that we may always hope—that there be no irrelevant exclusiveness from participation in its highest work, its powers, and its discourse. Ours, for special reasons of history, rendered more and more acute by the nature of the twentieth-century world, is a fluid society, with rapid change its hallmark. Like so many others, it is, in its politics, and much of its public life, a largely, even an inherently, secular society. We live, as we all know, with an expansion of knowledge overpoweringly beautiful, vast, ramified, quite unparalleled in the history of men. We live with a yearly enrichment of our understanding of nature, and of man as part of nature, that doubles every decade; and that is in its nature, necessarily, inevitably, and even in part happily an enrichment of specialization. This age of ours is the scientific age, in which our work, our leisure, our economy, and an increasingly large part of the very quality of our lives, are based on the application of newly acquired knowledge of nature to practical human problems. Size, egalitarianism, flux, are the social hallmarks of a continuing cognitive revolution. [. . .]

Living in today's world is not easy. No human society has ever solved problems that now confront us, or has even lived with them in dignity. This is for us not so much a time of anger as of honest sorrow, of renewal, of effort. Yet there are also great virtues in today's world: the recession of prejudice, of poverty, disease, and degradation; the creative, intimate, and lovely communities which thrive; the brilliance and wonder of the sciences. [. . .]

If I cannot be comforted by Vishnu's argument to Arjuna, it is because I am too much a Jew, much too much a Christian, much too much a European, far too much an American. For I believe in the meaningfulness of human history, and of our role in it, and above all of our responsibility to it.

Great cultures have flourished without this belief; perhaps they will again. If the switches of great war are thrown, in anger or in error, and if indeed there are human survivors, there may some-day again be high art, perhaps, and some ennobling sense of the place of man and his destiny. There may even be great science, but there will be no sense of history. There will be no sense of "progress in freedom."

This belief in progress and dedication to this belief have brought us where we are. All high civilizations have had a tradition of learning the truth, of contemplation, of under-standing. Since Greek times, many have understood as well the role of rigor, of proof, of anchoring consequence to hypothesis. They have had as well the art of putting questions to nature, of experiment; they have had forms of communication, perhaps inadequate, but at once robust and intimate. It has taken all these, rediscovered and slowly recaptured in the last millenium, to make the age of science; but it has taken more. Transfused with these, there has been a special sense of progress, not merely in man's understanding, but in the conditions of man's life, in his civility, in the nobility of his institutions and his freedom, a sense of progress not for the individual soul alone, but of progress in history, in man's long story.

We may well have learned that if we of the West do not look to our own virtue, and that of our institutions and our life and lives, we shall be ill-equipped to bring liberty to our colleagues now deprived of it, or to make either our culture or our liberty relevant and helpful to the lands newly embarked on unprecedented change. Let us in many varied ways turn to this, quite without flattery or illusion, but not quite without hope.

In the late fifties. (OMC)

10

ON SCIENCE AND CULTURE

1962

WE LIVE IN AN UNUSUAL WORLD, marked by very great and irreversible changes that occur within the span of a man's life. We live in a time where our knowledge and understanding of the world of nature grows wider and deeper at an unparalleled rate; and where the problems of applying this knowledge to man's needs and hopes are new, and only a little illuminated by our past history.

Indeed it has always, in traditional societies, been the great function of culture to keep things rather stable, quiet, and unchanging. It has been the function of tradition to assimilate one epoch to another, one episode to another, even one year to another. It has been the function of culture to bring out meaning, by pointing to the constant or recurrent traits of human life, which in easier days one talked about as the eternal verities.

In the most primitive societies, if one believes the anthropologists, the principal function of ritual, religion, of culture is, in fact, almost to stop change. It is to provide for the social organism what life provides in such a magic way for living organisms, a kind of homeostasis, an ability to remain intact, to respond only very little to the obvious convulsions and alterations in the world around.

Today, culture and tradition have assumed a very different intellectual and social purpose. The principal function of the most vital and living traditions today is precisely to provide the instruments of rapid change. There are many things which go together to bring about this alteration in man's life; but probably the decisive one is science itself. I will use that word as broadly as I know, meaning the natural sciences, meaning the historical sciences, meaning all those matters on which men can converse objectively with each other. I shall not continually repeat the distinction between science as an effort to find out about the world and understand it, on the one hand, and science, in its applications in technology, as an effort to do something useful with the knowledge so acquired. But certain care is called for, because, if we call this the scientific age, we make more than one kind of oversimplification. When we talk about science today, we are likely to think of the biologist with his microscope or the physicist with his cyclotron; but almost certainly a great deal that is not now the subject of successful study will later come to be. I think we probably today have under cultivation only a small part of the terrain which will be natural for the sciences a century from now. I think of the enormously rapid growth in many parts of biology, and of the fact, ominous but not without hope, that man is a part of nature and very open to study.

The reason for this great change from a slowly moving, almost static world, to the world we live in, is the cumulative character, the firmness, the givenness of what has been learned about nature. It is true that it is transcended when one goes into other parts of experience. What is true on the scale of the inch and the centimeter may not be true on the scale of a billion light-years; it may not be true either on the scale of a one-hundred billionth of a centimeter; but it stays true where it was proven. It is fixed. Thus everything that is found out is added to what was known before, enriches it, and does not have to be done over again. This essentially cumulative irreversible character of learning things is the hallmark of science.

This means that in man's history the sciences make changes which cannot be wished away and cannot be undone. Let me give two quite different examples. There is much talk about getting rid of atomic bombs. I like that talk; but we must not fool ourselves. The world will not be the same, no matter what we do with atomic bombs, because the knowledge of how to make them cannot be exorcised. It is there; and all our arrangements for living in a new age must bear in mind its omnipresent virtual presence, and the fact that one cannot change that. A different example: we can never have again the delusions about the centrality and importance of our physical habitat, now that we know something of where the earth is in the solar system, and know that there are hundreds of billions of suns in our galaxy, and hundreds of billions of galaxies within reach of the great telescopes of the world. We can never again base the dignity of man's life on the special character in space and time of the place where he happens to live.

These are irreversible changes; so it is that the cumulative character gives a paradigm of something which is, in other respects, very much more subject to question: the idea of human progress. One cannot doubt that in the sciences the direction of growth is progress. This is true of the knowledge of fact, the understanding of nature, and the knowledge of skill, of technology, of learning how to do things. When one applies this to the human situation, and complains that we make great progress in automation and computing and space research but no comparable moral progress, this involves a total misunderstanding of the difference between the two kinds of progress. I do not mean that moral progress is impossible; but it is not, in any sense, automatic. Moral regress, as we have seen in our day, is just as possible. Scientific regress is not compatible with the continued practice of science.

It is, of course, true, and we pride ourselves on it that it is true, that science is quite international, and is the same (with minor differences of emphasis) in Japan, France, the United States, Russia. But culture is not international; indeed I am one of those who hope that, in a certain sense, it never quite will be, that the influence of our past, of our history, which is for different reasons and different peoples quite different, will make itself felt and not be lost in total homogeneity.

I cannot subscribe to the view that science and culture are co-extensive, that they are the same thing with different names; and I cannot subscribe to the view that science is something useful, but essentially unrelated to culture. I think that we live in a time which has few historical parallels, that there are practical problems of human institutions, their obsolescence and their inadequacy, problems of the mind and spirit which, if not more difficult than ever before, are different, and difficult. I shall be dealing with some traits of the sciences which contribute to the difficulty, and may here give a synopsis of what they are. They have to do with the question of why the scientific revolution happened when it did; with the characteristic growth of the sciences: with their characteristic internal structure: with the relation of discovery in the sciences to the general ideas of man in matters which are not precisely related to the sciences: with freedom and necessity in the sciences, and the question of the creative and the open character of science, its infinity: and with what direction we might try to follow in bringing coherence and order to our cultural life, in doing what it is proper for a group of intellectuals, of artists, of philosophers, teachers, scientists, statesmen to do to help refashion the sensibility and the institutions of this world, which need refashioning if we are at all to survive.

It is not a simple question to answer why the scientific revolution occurred when it did. It started, as all serious historians would agree, in the late Middle Ages and early Renaissance, and was very slow at first. No great culture has been free of curiosity and reflection, of contemplation and thought. "To know the causes of things" is something that serious men have always wanted, a quest that serious societies have sustained. No great culture has been free of inventive genius. If we think of the culture of Greece, and the following Hellenistic and Roman period, it is particularly puzzling that the scientific revolution did not occur then. The Greeks discovered something without which our contemporary world would not be what it is: standards of rigor, the idea of proof, the idea of logical necessity, the idea that one thing implies another. Without that, science is very nearly impossible, for unless there is a quasi-rigid structure of implication and necessity, then if something turns out not to be what one expected, one will have no way of finding out where the wrong point is: one has no way of correcting himself, of finding the error. But this is something that the Greeks had very early in their history. They were curious and inventive; they did not experiment in the scale of modern days, but they did many experiments; they had as we have only recently learned to appreciate, a very high degree of technical and technological sophistication. They could make very subtle and complicated instruments; and they did, though they did not write much about it. Possibly the Greeks did not make the scientific revolution because of some flaw in communication. They were a small society, and it may be that there were not quite enough people involved.

In a matter of history, we cannot assign a unique cause, precisely because the event itself is unique; you cannot test, to see if you have it right. I think that the best guess is that it took something that was not present in Chinese civilization, that was wholly absent in Indian civilization, and absent also from Greco-Roman civilization. It needed an idea of progress, not limited to better understanding for this idea the Greeks had. It took an idea of progress which has more to do with the human condition, which is well expressed by the second half of the famous Christian dichotomy—faith and works; the notion that the betterment of man's condition, his civility, had meaning; that we all had a responsibility to it, a duty to it, and to man. I think that it was when this basic idea of man's condition, which supplements the other worldly aspects of religion, was fortified and fructified between the thirteenth and fifteenth centuries by the rediscovery of the ancient world's scientists, philosophers, and mathematicians, that there was the beginning of the scientific age. By the seventeenth century there were a handful of men involved in improving human knowledge, or "useful knowledge" as the phrase went, so that new societies like the Royal Society and the Academy were formed, where people could talk to each other and bring to the prosecution of science that indispensable element of working together, of communication, or correcting the other fellow's errors and admiring the other fellow's skills, thus creating the first truly scientific communities.

Just before Newton, Hobbes wrote:

> The Sciences are small power; because not eminent; and therefore, not acknowledged in any man, nor one at all, but in a few; and in them, but of a few things. For Science is of that nature, as none can understand it to be, but such as is good measure have attayned it.
>
> Arts of publique use, as Fortification, making of Engines, and other Instruments of War; because they conferre to Defense, and Victory, are Power.

It was the next century that put science in a context of fraternity, even of universal brotherhood. It encouraged a political view which was egalitarian, permissive, pluralistic, liberal—everything for which the word "democratic" is today justly and rightly used. The result is that the scientific world of today is also a very large

127

one: an open world in which, of course, not everybody does everything, in which not everybody is a scientist or a prime minister, but in which we fight very hard against arbitrary exclusion of people from any works, any deliberation, any discourse, any responsibility for which their talents and their interests suit them. The result is that we face our new problems, created by the practical consequences of technology, and the great intellectual consequences of science itself, in the context of a world of two or three billion people, an enormous society for which human institutions were not really ever designed. We are facing a world in which growth is characteristic, not just of the sciences themselves, but of the economy, of technology, of all human institutions; no one can open a daily paper without seeing the consequences.

One can measure scientific growth in a number of ways, but it is important not to mistake things. The excellence of the individual scientist does not change much with time. His knowledge and his power does, but not the high quality that makes him great. We do not look to anyone to be better than Kepler or Newton, any more than we look to anyone to be better than Sophocles, or to any doctrine to be better than the gospel according to St. Matthew. Yet one can measure things, and it has been done. One can measure how many people work on scientific questions: one can count them. One can notice how much is published.

These two criteria show a doubling of scientific knowledge in every ten years. Casimir calculated that if the *Physical Review* continued to grow as rapidly as it has between 1945 and 1960, it would weigh more than the earth during the next century. In fifteen years, the volume of chemical abstracts has quadrupled; in biology the changes are faster still. Today, if you talk about scientists and mean by that people who have devoted their lives to the acquisition and application of new knowledge, then 93 per cent of us are still alive. This enormously rapid growth, sustained over two centuries, means, of course, that no man learned as a boy more than a small fraction in his own field of what he ought to know as a grown man.

There are several points to keep in mind. One would naturally think that if we are publishing so much, it must be trivial. I think that this is not true: any scientific community with sane people would protect itself against that: because we have to read what is published. The argument not to permit the accumulation of trivial, unimportant things which are not really new, which do not add to what was known before, is overwhelming.

The second point is that one may say that every new thing renders what was known before uninteresting, that one can forget as rapidly as one learns. That is in part true: whenever there is a great new understanding, a great new element of order, a new theory, or a new law of nature, then much that before had to be remembered in isolation becomes connected and becomes, to some extent, implied and simplified. Yet one cannot forget what went before, because usually the meaning of what is discovered in 1962 is to be found in terms of things that were discovered in 1955 or 1950 or earlier. These are the things in terms of which the new discoveries are made, the origins of the instruments that give us the new discoveries, the origins of the concepts in terms of which they are discovered, the origins of the language and the tradition.

A third point: if one looks to the future of something that doubles every ten years, there must come a time when it stops, just as the *Physical Review* cannot weigh more than the earth. We know that this will saturate, and probably at a level very much higher than today; there will come a time when the rate of growth of science is not such that in every ten years the amount that is known is doubled; but the amount that is added to knowledge then will be far greater than it is today. For this rate of growth suggests that, just as the professional must, if he is to remain professional, live a life of continuous study, so we may find a clue here also to the more general behavior of the intellectual with regard to his own affairs, and those of his colleagues in somewhat different fields. In the most practical way a man will have some choice: he may choose to continue to learn about his own field in an intimate, detailed, knowledgeable way, so that he knows what there is to know about it. But then the field will not be very wide. His knowledge will be highly partial of science as a whole, but very intimate and very complete of his own field. He may, on the other hand, choose to know generally, superficially, a good deal about what goes on in science, but without competence, without mastery, without intimacy, without depth. The reason for emphasizing this is that the cultural values of the life of science almost all lie in the intimate view: here are the new techniques, the hard lessons, the real choices, the great disappointments, the great discoveries.

All sciences grow out of common sense, out of curiosity, observation, reflection. One starts by refining one's observation and one's words, and by exploring and pushing things a little further than they occur in ordinary life. In this novelty there are surprises; one revises the way one thinks about things to accommodate the surprises; then the old way of thinking gets to be so cumbersome and inappropriate that one realizes that there is a big change called for, and one recreates one's way of thinking about this part of nature.

Through all this one learns to say what one has done, what one has found, and to be patient and wait for others to see if they find the same things, and to reduce, to the point where it really makes no further difference, the normally overpoweringly vital element of ambiguity in human speech. We live by being ambiguous, by not settling things because they do not have to be settled, by suggesting more than one thing because their co-presence in the mind may be a source of beauty. But in talking about science one may be as ambiguous as ever until we come to the heart of it. Then we tell a fellow just what we did in terms that are intelligible to him, because he has been schooled to understand them, and we tell him just what we found and just how we did it. If he does not understand us, we go to visit him and help him; and if he still does not understand us, we go back home and do it over again. This is the way in which the firmness and solidity of science is established.

How then does it go? In studying the different parts of nature, one explores with different instruments, explores different objects, and one gets a branching of what at one time had been common talk, common sense. Each branch develops new instruments, ideas, words suitable for describing that part of the world of nature. This treelike structure, all growing from the common trunk of man's common primordial experience, has branches no longer associated with the same question, nor the same words and techniques. The unity of science, apart from the fact that it all has a common origin in man's ordinary life, if not a unity of deriving one part from another, nor of finding an identity between one part and another, between let us say, genetics and topology, to take two impossible examples, where there is indeed some connection.

The unity consists of two things: first and ever more strikingly, an absence of inconsistency. Thus we may talk of life in terms of purpose and adaptation and function, but we have found in living things no tricks played upon the laws of physics and chemistry. We have found and I expect will find a total consistency, and between the different subjects, even as remote as genetics and topology, an occasional sharp mutual relevance. They throw light on each other; they have something to do with each other: often the greatest things in the sciences occur when two different discoveries made in different worlds turn out to have so much in common that they are examples of a still greater discovery.

The image is not that of an ordered array of facts in which every one follows somehow from a more fundamental one. It is rather that of a living thing: a tree doing something that trees do not normally do, occasionally having the branches grow together and part again in a great network.

The knowledge that is being increased in this extraordinary way is inherently and inevitably very specialised. It is different for the physicist, the astronomer, the microbiologist, the mathematician. There are connections: there is this often important mutual relevance. Even in physics, where we fight very hard to keep the different parts of our subject from flying apart (so that one fellow will know one thing and another fellow will know another, and they do not talk to each other), we do not entirely succeed, in spite of a passion for unity which is very strong. The traditions of science are specialized traditions; this is their strength. Their strength is that they use the words, the machinery, the concepts, the theories, that fit their subjects; they are not encumbered by having to try to fit other sorts of things. It is the specialized traditions which give the enormous thrust and power to the scientific experience. This also makes for the problem of teaching and explaining the sciences. When we get to some very powerful general result which illuminates a large part of the world of nature, it is by virtue of its being general in the logical sense, of encompassing an enormous amount of experience in its concepts; and in its terminology it is most highly specialized, almost unintelligible except to the men who have worked in the field. The great laws of physics today, which do not describe everything (or we would be out of business) but which underlie almost everything that is ever noticed in ordinary human experience about the physical world, cannot be formulated in terms that can reasonably be defined without a long period of careful schooling. This is comparably true in other subjects.

One has then in these specializations the professional communities in the various sciences. They are very intimate, work closely together, know each other throughout the world. They are always excited—sometimes jealous but usually pleased—when one member of the community makes a discovery. I think, for instance, that what we now call psychology will one day perhaps be many sciences, that there will be many different specialized communities practicing them, who will talk with one another, each in their own profession and in their own way.

These specialized communities, or guilds, are a very moving experience for those who participate. There have been many temptations to see analogues in them for other human activities. One that we hear much discussed is this: "If physicists can work together in countries with different cultures, in countries with different politics, in countries of different religions, even in countries which are politically obviously hostile, is not this a way to bring the world together?"

The specializing habits of the sciences have, to some extent, because of the tricks of universities, been carried over to other work, to philosophy and to the arts. There is technical philosophy which is philosophy as a craft, philosophy for other philosophers, and there is art for the artists and the critics. To my mind, whatever virtues the works have for sharpening professional tools, they are profound misreadings, even profound subversions of the true functions of philosophy and art, which are to address themselves to the general common human problem. Not to everybody, but to anybody: not to specialists.

It is clear that one is faced here with formidable problems of communication, of telling people about things. It is an immense job of teaching on all levels, in every sense of the word, never ending.

It has often been held that the great discoveries in science, coming into the lives of men, affect their attitudes toward their place in life, their views, their philosophy. There is surely some truth in this.*

If discoveries in science are to have an honest effect on human thought and on culture, they have to be understandable. That is likely to be true only in the early period of a science, when it is talking about things which are not too remote from ordinary experience. Some of the great discoveries of this century go under the names of Relativity and Uncertainty, and when we hear these words we may think, "This is the way I felt this morning: I was relatively confused and quite uncertain": this is not at all a notion of what technical points are involved in these great discoveries, or what lessons.

I think that the reason why Darwin's hypothesis had such an impact was, in part, because it was a very simple thing in terms of ordinary life. We cannot talk about the contemporary discovery in biology in such language, or by referring only to things that we have all experienced.

Thus I think that the great effects of the sciences in stimulating and in enriching philosophical life and cultural interests have been necessarily confined to the rather early times in the development of a science. There is another qualification. Discoveries will really only resonate and change the thinking of men when they feed some hope, some need that preexists in the society. I think that the real sources of the Enlightenment, fed a little by the scientific events of the time, came in the rediscovery of the classics, of classic political theory, perhaps most of all the Stoics. The hunger of the eighteenth century to believe in the power of reason, to wish to throw off authority, to wish to secularize, to take an optimistic view of man's condition, seized on Newton and his discoveries as an illustration of something which was already deeply believed in quite apart from the law of gravity and the laws of motion. The hunger with which the nineteenth

---

*Examples that are usually given include Newton and Darwin. Newton is not a very good example, for when we look at it closely we are struck by the fact that in the sense of the Enlightenment, the sense of a coupling of faith in scientific progress and man's reason with a belief in political progress and the secularization of human life, Newton himself was in no way a Newtonian. His successors were.

century seized on Darwin had very much to do with the increasing awareness of history and change, with the great desire to naturalize man, to put him into the world of nature, which existed long before Darwin and which made him welcome. I have seen an example in this century where the great Danish physicist Niels Bohr found in the quantum theory when it was developed thirty years ago this remarkable trait: it is consistent with describing an atomic system, only much less completely than we can describe large-scale objects. We have a certain choice as to which traits of the atomic system we wish to study and measure and which to let go; but we have not the option of doing them all. This situation, which we all recognize, sustained in Bohr his long-held view of the human condition: that there are mutually exclusive ways of using our words, our minds, our souls, any one of which is open to us, but which cannot be combined: ways as different, for instance, as preparing to act and entering into an introspective search for the reasons for action. This discovery has not, I think, penetrated into general cultural life. I wish it had; it is a good example of something that would be relevant, if only it could be understood.

Einstein once said that a physical theory was not determined by the facts of nature, but was a free invention of the human mind. This raises the question of how necessary is the content of science—how much is it something that we are free not to find—how much is it something that could be otherwise? This is, of course, relevant to the question of how we may use the words "objectivity" and "truth." Do we, when we find something, "invent" it or "discover" it?

The fact is, of course, just what one would guess. We are, of course, free in our tradition and in our practice, and to a much more limited extent individually, to decide where to look at nature, and how to look at nature, what questions to put, with what instruments and with what purpose. But we are not the least bit free to settle what we find. Man must certainly be free to invent the idea of mass, as Newton did and as it has been refined and redefined; but having done so, we have not been free to find that the mass of the light quantum or the neutrino is anything but zero. We are free in the start of things. We are free as to how to go about it; but then the rock of what the world is, shapes this freedom with a necessary answer. That is why ontological interpretations of the word "objective" have seemed useless, and why we use the word to describe the clarity, the lack of ambiguity, the effectiveness of the way we can tell each other about what we have found.

Thus in the sciences, total statements like those that involve the word "all," with no qualifications, are hardly ever likely to occur. In every investigation and extension of knowledge we are involved in an action; in every action we are involved in a choice; and in every choice we are involved in a loss, the loss of that we did not do. We find this in the simplest situations. We find this in perception, where the possibility of perceiving is coextensive with our ignoring many things that are going on. We find it in speech where the possibility of understandable speech lies in paying no attention to a great deal that is in the air, among the sound waves, in the general scene. Meaning is always attained at the cost of leaving things out. We find it is the idea of complementarity here in a sharp form as a recognition that the attempt to make one sort of observation on an atomic system forecloses others. We have freedom of choice, but we have no escape from the fact that doing some things must leave out others.

In practical terms, this means, of course, that our knowledge is finite and never all-encompassing. There is always much that we miss, much that we cannot be aware of because the very act of learning, of ordering, of finding unity and meaning, the very power to talk about things means that we leave out a great deal.

Ask the question: *Would another civilization based on life on another planet very similar to ours in its ability to sustain life have the same physics?* One has no idea whether they would have the same physics or not. We might be talking about quite different questions. This makes ours an open world without end. I had a Sanskritist friend in California who used to say mockingly that, if science were any good, it should be much easier to be an educated man now than it was a generation ago. That is because he thought the world was closed.

The things that make us choose one set of questions, one branch of inquiry rather than another are embodied in scientific traditions. In developed sciences each man has only a limited sense of freedom to shape or alter them; but they are not themselves wholly determined by the findings of science. They are largely of an aesthetic character. The words that we use: simplicity, elegance, beauty, indicate that what we grope for is not only more

135

knowledge, but knowledge that has order and harmony in it, and continuity with the past. Like all poor fellows, we want to find something new, but not something too new. It is when we fail in that, that the great discoveries follow.

All these themes—the origin of science, its pattern of growth, its branching reticular structure, its increasing alienation from the common understanding of man, its freedom, the character of its objectivity and its openness—are relevant to the relations of science and culture. I believe that they can be and should be far more robust, intimate, and fruitful than they are today.

I am not here thinking of the popular subject of "mass culture." In broaching that, it seems to me one must be critical but one must, above all, be human; one must not be a snob; one must be rather tolerant and almost loving. It is a new problem; one must not expect it to be solved with the methods of Periclean Athens. In the problems of mass culture and, above all, of the mass media, it is not primarily a question of the absence of excellence. The modest worker, in Europe or in America, has within reach probably better music and more good music, more good art, more good writing than his predecessors have ever had. It seems rather that the good things are lost in such a stream of poor things, that the noise level is so high, that some of the conditions for appreciating excellence are not present. One does not eat well unless one is hungry; there is a certain frugality to the best cooking; and something of this sort is wrong with the mass media. But that is not now my problem.

Rather, I think loosely of what we may call the intellectual community: artists, philosophers, statesmen, teachers, men of most professions, prophets, scientists. This is an open group, with no sharp lines separating those that think themselves of it. It is a growing faction of all peoples. In it is vested the great duty for enlarging, preserving, and transmitting our knowledge and skills, and indeed our understanding of the interrelations, priorities, commitments, injunctions, that help men deal with their joys, temptations, and sorrows, their finiteness, their beauty. Some of this has to do, as the sciences so largely do, with propositional truth, with propositions which say "If you do thus and so you will see this and that"; these are objective and can be checked and crosschecked; though it is always wise from time to time to doubt, there are ways to put an end to the doubt. This is how it is with the sciences.

In this community there are other statements which "emphasize a theme" rather than declare a fact. They may be statements of connectedness or relatedness or importance, or they may be in one way or another statements of commitment. For them the word "certitude," which is a natural norm to apply in the sciences, is not very sensible—depth, firmness, universality, perhaps more—but certitude, which applies really to verification, is not the great criterion in most of the work of a philosopher, a painter, a poet, or a playwright. For these are not, in the sense I have outlined, objective. Yet for any true community, for any society worthy of the name, they must have an element of community of being common, of being public, of being relevant and meaningful to man, not necessarily to everybody, but surely not just to specialists.

I have been much concerned that, in this world of change and scientific growth, we have so largely lost the ability to talk with one another, to increase and enrich our common culture and understanding. And so it is that the public sector of our lives, what we hold and have in common, has suffered, as have the illumination of the arts, the deepening of justice and virtue, and the ennobling power of our common discourse. We are less men for this. Never in man's history have the specialized traditions more flourished than today. We have our private beauties. But in those high undertakings when man derives strength and insight from public excellence, we have been impoverished. We hunger for nobility, the rare words and acts that harmonize simplicity with truth. In this default I see some connection with the great unresolved public problems—survival, liberty, fraternity.

In this default I see the responsibility that the intellectual community has to history and to our fellows: a responsibility which is a necessary condition for remaking human institutions as they need to be remade today that there may be peace, that they may embody more fully those ethical commitments without which we cannot properly live as men.

This may mean for the intellectual community a very much greater effort than in the past. The community will grow; but I think that also the quality and the excellence of what we do must grow. I think, in fact, that with the growing wealth of the world, and the possibility that it will not all be used to make new committees, there may indeed be genuine leisure, and that a high commitment on this leisure is that we reknit the discourse and the understanding between the members of our community.

In this I think we have, all of us, to preserve our competence in our own professions, to preserve what we know intimately, to preserve our mastery. That is, in fact, our only anchor in honesty. We need also to be open to other and complementary lives, not intimidated by them and not contemptuous of them (as so many are today of the natural and mathematical sciences). As a start, we must learn again, without contempt and with great patience, to talk to one another; and we must hear.

Robert Oppenheimer with President Johnson at the Fermi Award
presentation in 1963. (OMC)

11

THE POWER TO ACT
The Scientific Revolution
and its Effects on
Democratic Institutions

1963

[...] FROM BACON AND JOHN DONNE through Henry Adams and Whitehead to now, wise men have foreseen, been frightened by, have applauded what is called the explosion of science. It is true that for over two hundred years there has been a constant growth in the number of people working on scientific problems and in the number of works they have made and published about our knowledge of nature, of ourselves as a part of nature, and on the applications of science in technology. Indeed, the scientific revolution has white hair; but it still has young eyes.

[. . .] This is a special time. We have had, as John Donne said, the destruction of an order of belief and of society. We have had, as Henry Adams said, a monstrous growth. We have had, as Whitehead said, changes that take place within the life of a man, even within a small part of the life of a man. We have today such a wealth of scientific knowledge that we may even think in terms of saturation. We may even think in elementary terms of arithmetic that the growth cannot accelerate for many more decades by the exponential law of growth characteristic of the past centuries. Never have the imminence and sharpness of change, never have the range and the vastness of our powers, never have the gravity and depth and beauty of the choices open to us, been greater.

[. . .] Democratic institutions and the scientific revolution have an historic association. Both were conceived in the Europe of the thirteenth century; both have continued in Europe and America, with a growing robustness, and with a fantastic increase of the

141

sciences, with a felt and understood connection of our knowledge and skill, on the one hand, and on the other of man's progress not merely in knowledge but in the civility of his life on earth. This did not happen in Greece or in China because there the ingredient of looking to human betterment was not so omnipresent. Perhaps this has to do with the differences in the Judeo-Christian traditions and the religions of the East. Certainly by the time of Franklin and Jefferson these relations were articulated.

It is, of course, not true that democratic institutions are needed for science to flourish. Thus, in the Japan of 1935, Yukawa made one of the great discoveries of physics. Today in the Soviet Union, quite unevenly, but quite seriously, great work in science is underway. The young people in China, though few of us know this at first hand, are, with great humility and great sense of purpose, building up their scientific tradition. But no matter in what culture it occurs, the nature of science is essentially a democratic one and we should not underestimate the contagion and the power and the strength of the tradition which we have inherited: for just this is the measure of our responsibility.

Although reading Jefferson today, we have a painful, nostalgic, and acute sense of the irrelevance of what he said, we do owe a surprising measure of our own democratic institutions to him, to his life, to his time. He expected that these institutions would change rather more than they have; he articulated the connection between a free society and the growth of knowledge and its application to the alleviation of illness, hunger, tyranny, and superstition. [. . .]

Jefferson was clear that, as the basis of all democratic institutions, citizens were informed, were responsible for their conduct, and shared, with all variations, some common sense of beauty and of virtue. It is in the words, *informed* and *responsible*, that we find our troubles. The problem of being informed, that is, the problem of adequate communication is today one which needs all our help. Communication is difficult because the things we took for granted are not true today; communication is difficult because the world is not only large, but accessible, so that it is on our minds and consciences, as it should be. One of the things we are learning is that the teaching needed in this time is, as it has always been, a difficult intellectual job; we cannot teach things only as they were learned, for that is too hard; nor can we teach them only as they apply, for that is too superficial. We must teach them for the truth that is in them, because it is that truth that is beautiful. We know that the job of teaching, the use of every kind of medium, the honest sharing of what is known and thought and loved, is a problem to which we must devote more and better resources. We know that in the world of the future what the French call permanent education is here to stay, and that no one will ever be through, or, rather, that no one need ever be through, and that many men will not be through with devoting a substantial part of their life, their time, their heart to learning more.

This ideal of an open, accessible world of knowledge, open and connected, is one which will have to be fought for. It is threatened by the cheap and the vulgar; it is threatened by all restrictions on freedom of communication, such as those that the Chinese Communist government and our own have collaborated to erect between our two countries; it is not a job for one luncheon or one hour. [. . .]The most dangerous errors are made when we imagine and act on the basis of what we might know, instead of knowing something and recognizing where it stops. [. . .]

In addition to knowledge, there is another requirement of democratic institutions, if they are to work, if they are to grow and change with the times: the power to act. The power to act is coupled with responsibility for actions. This possibility of action is never complete, never always exercised rightly; yet it must be very widespread, and in terms of laws and institutions, potentially quite universal. Let us look at two current constellations of questions: peace and social justice.

[. . .] We all know that just in this country the colored people, the permanently unemployed, the young, the ill, the old, the men and women living as prisoners in their ugly cities, all are ill cared for by our institutions and our practices, which have been made evil just by the powers derived from science. They could all be remedied. It is clear that our institutions are not nearly adequate to receive the great resources of human warmth and dedication immanent in our young people. It is clear that our institutions are not yet adequate to let men and women do what they would to play a responsible part in the betterment and ennobling and beauty of our society. That is one reason why we abdicate to such a dismal extent to the one thing whose institutional basis is hypertrophic in our society: the dollar.

There is another aspect of social justice even greater than our injustices at home: the injustice between our land and other lands and peoples. We all know with what care such matters must be touched, because we are dealing here not just with the creation of economies, constitutions, and universities. We are in the beginning and in the end dealing with people, who must form their lives, in some way, in the light of the awesome example of European society, yet in their own interests, from where they are and not from where we might think they ought to be. Though it is hard, though our record shows much bad as well as good, we cannot ignore or reject the disparity between the poor and the rich without corrupting our own life, without endangering its rightness. We cannot ignore them without that danger, because of the problem of peace and the problem of war. Here most of us feel powerless; most of us, only by a certain rugged, unyielding tension, are aware of our responsibility to the whole future of man.

In these times, in these years, the atom bomb and nuclear weapons preside over our anxieties. This is an accident. It was, of course, done by design, but it was an accident that it could have been done when it was. The knowledge that we find of nature, whether it is of life, or of ourselves, or of chemistry, this knowledge could have produced, has produced, may yet produce very terrible instruments of major war different than the bombs, perhaps not more terrible, but just as ineluctable. These instruments will not really go away; I hope they will be dealt with. In a society which remembers what it has learned and how to do things that it has done, which is incredibly rich in potential and in fact, these weapons are as present as the desire to have them and to use them. We can only hope that they will increasingly appear irrelevant and thus in the end preposterous, that some day we will look back ashamed of how stupid we were. These are problems of all men and must be dealt with through an openness of communication to all men, an openness which is democratic and can only be promoted by an extension of democratic institutions [. . .].

In 1946 we tried a little in sketching a plan for the control of atomic energy. We made a pilot suggestion of how that might go, stressing a deep community of knowledge, and common constructive goals, as the right framework for averting the dangers of

144

atomic warfare. That was, of course, very narrow, yet it was categorically rejected by the Soviet Union and from talk with the delegates of the United Kingdom and France, it was clear that they had no great love for this affair, which would get them all mixed up with people with whom they did not much want to talk or work. Even the United States, which was never called on to see if it would fish or cut bait, would, as the published comments on Senatorial resistance to a ban on tests of atomic weapons suggest, have had very rough going if we had come to try to make an international community to deal with broad aspects of the atomic problem, and thus the problem of growing knowledge and power.

What we do have is an increasingly wide and deep under-standing of the terror and horror and wrongness of war, of the irreversibility of the powers acquired by our knowledge of nature. We have one other resource, very slow to come and that is a recognition that there are in our common tradition some things that are relevant to this predicament: for one, our responsibility as men alone and as men banded together for the future, not just as something good for ourselves, or to preserve and love the past, but to preserve and cherish the future; for another, our under-standing that evil is the monopoly of no people, that we can and must see it in ourselves and even in our own country. For in this creation of new institutions on a large, perhaps even world-wide scale, we have the great problem of politics in its universal and most acute form, to reconcile our detestation of evil as we find it abroad and our love of those things we cherish, and to reconcile both of these with justice. This is the hardest and highest achievement of politics, surely as demanding, as difficult, as the creation of the institutions with which our own country has been blessed from the beginning. More than in the nuclear weapons, more than in the increased ability to cure or alleviate disease, more than in the revolutions in communications, travel, automa-tion, in abundant food and power, it is in this, and what it recalls to us of the strength and relevance of our tradition to a new time, that lies my hope, lies our hope, for the impact of the scientific revolution on the sources of democratic institutions.

Feeding son Peter, around 1942. (OMC)

12

A WORLD WITHOUT WAR

1963

[. . .] IN AN IMPORTANT SENSE, the sciences have solved the problem of communicating with one another more completely than has any human enterprise. To retell an old story, thirty-five years ago, Dirac and I were in Göttingen. He was developing the quantum theory of radiation, and I was a student. He learned that I sometimes wrote a poem, and he took me to task, saying, "In physics we try to say things that no one knew before in a way that everyone can understand, whereas in poetry . . . "

It is one of our old and consistent traditions to be concerned with the words we use, and with their purification, and thus with the concepts with which we describe nature. It was true of Newton, of Lavoisier, of Cauchy, of Mendel, and of course, in our day, of Einstein and of Bohr. [. . .]

When we tell about our work, we explain what we have done and we tell what we have seen, whether we are describing a radioastronomical object, a new property of fiber bundles, or the behavior of men attempting to solve problems. We are prepared to believe that the explicit content of science has its roots in these accounts of action, which are often factual, but often fore-shortened and synoptic, because they are cast in terms which scientific traditions have established long ago.

Among us there is surely a great and appropriate variation in how we describe this foundation for the objectivity of our knowledge, and for the lack of ambiguity in the terms we use to tell of it. Of course, there is an even wider latitude in what we think of the reasons for the success of science, in what attributes of the world of nature in which we find ourselves underlie the manifestations of order which are our business. Why can we work on the same table and with the same test tube when we cannot have the same melancholy or the same resolution? Why does so much of the order of the natural world find its expression in number and more abstract mathematical structure?

[. . .] The foundation for scientific knowledge precludes much that is an essential part of man's life. One cannot be a very effective scientist if he is a practicing solipsist. We cannot expect to describe a common world of introspection by telling people what we have done and what we have seen, though we probably can, and increasingly will, describe elements of behavior which may have some correspondence to our inner world. Among these things of which we cannot talk without some ambiguity, and in which the objective structure of the sciences will play what is often a very minor part, but sometimes an essential one, are many questions which are not private, but which are common questions, and public ones: questions of the arts, the good life, the good society. There is no reason why we should come to these with a greater consensus or a greater sense of valid relevant experience than any other profession. They need reason, and they need a preoccupation with consistency, but only insofar as the scientist's life has analogies with the artist's—and in important ways it does—only insofar as the scientist's life is in some way a good life, and his society a good society, have we any professional credentials to enter these discussions, and not primarily because of the objectivity of our communication and our knowledge. It is doubtful that we have a special qualification for these matters, and even more doubtful that our professional practices should disqualify us, or that we should lose interest and heart in preoccupations which have ennobled and purified men throughout history, and for which the world has great need today.

[. . .] The constant concern within the scientific enterprise to purify and refine our language is, of course, a sort of parody of what we are all about. Scientists do not really do this except in moments of crisis, or in order to make way for something very new and deep. We come to our new problems full of old ideas and old words, not only the inevitable words of daily life, but those which experience has shown fruitful over the years. This is an inevitable approach to the new; when it is not too new, it gets by.

The understanding of scientific knowledge, however, is a very different thing from being the recipient of a communication. There is an element of action inseparable from understanding: to question, to try, to apply, to adapt, to ask new questions, to see if one understands, and to test what one has been told. We need, at times, to talk about the sources and the springs of this action, without which communication would provide the fuel pipes, the electrical wiring, the transmission of a car, but not the combustion which gives it power and life.

We do not talk of this very well: imagination, play, curiosity, invention, action, these are all involved. They are indeed only rarely all combined, and supplemented by skepticism and criticism, in any one man in any one moment; one of the charms of the scientific enterprise is how deficient we can be in many of these qualities and still play some meaningful part in it.

We know that we love the old words, the old imagery, and the old analogies. We keep them for more and more unfamiliar and more and more unrecognizable things. Think of "wave," "information," "relativity." We know that one can explore and study the springs of the movement of science, that it is a fit if very difficult subject of study. Today at least we are not able to talk about it very well, not at all as well as we can talk of molecules or galaxies, or even of the effective definition of the words that we use. Yet without a living engagement there is no understanding and there is no life of science. We know that we cannot command this, or perhaps even learn it, except by apprenticeship, by following what others have done, and by listening to the mischievous voices of adventure and play and exploration and doubt with which we greet a new experience or a new communication. This has very much to do with what we can in practice and honesty mean by the unity of science. Think, for instance, of contemporary mathematics. Up to our time, it has been the experience of our enterprise that there have been a good number of men who combined creation and wide knowledge of the mathematics of their day with a lively interest in those elements of the natural sciences in which this mathematical order might be embodied. This conversation, as a lively mutual understanding, is rather thin today. It is not rare to find a physical scientist who will hear of some beautiful new result—in algebra, for instance, or topology—with pleasure, with amazement, and with admiration.

149

It is unlikely, though, that he will be deeply engaged by it and try to see how it affects other things he may have known, or thought to know. It is also true that many mathematicians will accept with a certain interest that there are in nature two neutrinos which have different properties, or that astronomers believe that they may be witnessing evidence of very massive gravitational implosion in other galaxies. It is good that we still do tell each other these pieces of news, but I would hope that the century-long tradition of a felt sense of reciprocal relevance between mathematics and natural science soon again find itself embodied in many of us, or, far more plausibly, in our successors.

Thus between us in the scientific professions, there is a partly accidental quality to the effectiveness of our converse with one another, and thus to the effective unity of our view of the world. There are two reasons for deploring this. One is that past experience suggests so strongly that among the sciences there are elements of relevance and mutual enlightenment which make such converse an essential part of deep and rapid progress; the other is that we regret for ourselves what we cannot really tell them. This is, of course, a reflection, within the internal society of the scientific enterprise, of a situation that characterizes our relations with human societies as a whole, with the society within which we are embedded, and that leaves us with problems, some very grave, and by no means all clearly soluble, having to do with the communication and comprehension of scientific knowledge. These problems rest on human weakness and limitation; but more specifically they rest on at least three features of the scientific enterprise which it has in common with the world in which its whole action takes place: size and saturation, growth and change, and specialization. [. . .]

Growth and change imply size, and growth and change are very deep in the nature of scientific enterprise. Without them we would not recognize the rooms in which we were living, or what our days were all about.

As for specialization, it is what sharpens our tools and our words, and is the instrument of penetrating deeper and farther into the world of nature.

We can and must live with these problems. Some of us will know one thing, some fewer will know many things, and the unity of our knowledge, its freedom from contradiction and its important and often very deep common relevance, will not preclude, but will be enriched by the great and blessed diversity of man.

The limits on the communication and comprehension of scientific knowledge which we find among ourselves have their analogies in the related but vaster limitations that we have in our external relations with those who are not yet, or not ever, involved in the scientific enterprise. The first of these is with the young, those who may be entering the life of science, and perhaps also, perhaps even more importantly, those who may not. [...] We have to express an appreciation to those who have been studying and practicing the teaching of the sciences in the schools and colleges, so that first sight shall not repel, and the institutions not resist the natural curiosity and love and joy of the experience, but open it, so that as many as can will have an opportunity to discover some trait of nature, to see with welcome some sure sign of order in nature, with their own hands and their own heads.

[...] We mostly take it for granted, though it is not quite obvious, that we would like to have this opening of the world of science, this induction into it, effective not only for those who will be of our company, but for as many as may be of all the young, and the newly young who are willing to study. There are two reasons why we hold this view. On the one hand we increasingly feel the need for companionship and for help. I am not speaking of the patronage of science, which has not been ungenerous in the past years, though it may come to be so. I have in mind rather what we all know, that more rapidly than ever before, the sciences have been embodied in new technologies, and that these bring on the scene new powers and possibilities, now a new need, now a new opportunity. These needs and opportunities often are relevant to what in us, and in most men, are the most deeply held convictions of what is right and good, convictions rooted in a long tradition, and integrally a part of our sensibility. Most of us are committed to preserving life and health where that is possible. Increasingly, and largely because of the effects during the last centuries of technology and industrialization on its modes, we are committed to limiting, if possible, to eliminating, war. We are committed to reducing labor and drudgery, and not only the hard labor of the field and the mine and the galley, but the dull labor of the factory. We are now clearly engaged in a great enterprise testing whether we can live in a world in which war does not play its traditional part, an enterprise in which not only long-inherited human institutions, but even older, even other more permanent human attitudes of anger, hatred, solidarity, self-importance, self-right-eousness, which war has fed, can permit the change. We are in this too deeply to let the good news or the bad news of the day or month or year affect or limit our hope and, where it is possible for us, our engagement in this great, open, unsettled action of man's history.

With the preservation of life too, and along with it the alteration and automation of work, we are concerned not only with the inadequacy of our institutions, which were framed for a very different world, but with our attitude toward the meaning and value and nature and quality of human life, so largely in our past built on productive work as its foundation. Here in this country we see the mixed fruits of medical and engineering technology first with the young and the old. It is reasonable to expect that they will spread, and that they will characterize many other technologically developed societies. I know of the concern that even the saving of children's lives may have created problems with which no one can cope, that have some bearing on the growth and size of the human society.

I do not suppose that a thorough knowledge of science, which is essentially unavailable to all of us, would really be helpful to our friends in other ways of life in acting with insight and courage in the contemporary world. Such a knowledge would perhaps be good in developing a greater recognition of the quality of our certitudes, where we are dealing with scientific knowledge that really exists, and the corresponding quality of hesitancy and doubt when we are assessing the probable course of events, the way in which men will choose and act, to ignore or to apply [. . .] the technological possibilities recently opened. Some honest and remembered experience of the exploration of nature, of dis-covery, and of the way in which we talk to one another about these things, might indeed be helpful. It would remove barriers and encourage an effective and trusting converse between us, and make more fruitful the indispensable role of friendship. [. . .] We have a modest part to play in history, and the barriers between us and the men of affairs, the statesmen, the artists, the lawyers, with whom we should be talking, could perhaps be markedly reduced if more of them knew a little of what we were up to, knew it with pleasure and some confidence; and if we were prepared to recognize both the important analogies between what moves us to act and to know, and the extraordinary and special quality of our experience and our communication about it with one another. [. . .].

The other reason for hoping that young people who will not be professional scientists, and older people who are young in heart, could have a greater scientific literacy and some limited experience, as ours also is limited, is that we know that the experience of scientific discovery is a good and beautiful experience, and an unforgettable one. We know that this is true even of little discoveries, and we understand that with the great ones it is shattering. It was on his seventy-first birthday that Einstein said to me, "When it has once been given a man to do some sensible things, afterwards his life is a little different." It seems not really as an act of arrogance but simply human, and not in the pejorative sense of the word, to wish these pleasures for as many of our fellows as can have them.

In our world, many things that men do rather naturally, that they have learned to do long, long ago, have become professions, have become part of the market. I think of song and sport and the arts, the practical arts and the fine arts. None of these is without discipline; and although they are very different from those that lead to the sciences, I would be slow to rate them easier. Yet people sing and make sport and practice the arts quite apart from the market, quite apart from a career. It would be a poorer, thinner life without that. Though surely we will not all burst into song, or take to skis, or pick up a chisel or a brush, some of us have done some of these things, and some of us will. It seems a proper hope that in our education, both for the young, and for those, in growing number, who like us have kept a lifelong taste for it, we do what we can to open the life of science at least as wide as that of song and the arts. Not everybody will want our pleasures, as among us not everyone can taste the others', as even we cannot expect an astronomer and a biologist fully to share what each has. We think of science as a very high and lovely part of life which, with all its discipline, is still directly responsive to a very

153

deep human need. We all know this, and all share it; but each of us, I think, must be free to use his own words to sing its praise, even to describe it.

We may be seeing a time in which war will come to play a smaller and an increasingly trivial part in man's life. We may be coming to a time in which, for growing parts of the world, the production of goods will require a much more minor commitment of human effort and life, and the market will leave men with a far greater measure of freedom. [. . .] For all this it will clearly not be enough that we preserve the integrity of our communication and comprehension, either among us, or with our fellows, but it is the least that we can do.

Robert Oppenheimer and General Groves at Ground Zero in 1945. (UPI Photo)

13

L'INTIME ET LE COMMUN—THE INTIMATE AND THE OPEN

1964

AS MY TITLE INDICATES, perhaps somewhat obscurely, I shall talk about one aspect of our life which may change very much in the times ahead, which has to do with how we will live tomorrow, "Comment vivre demain." [. . .] I am aware, as we all are, but perhaps a little more sharply, that there is still an open question of whether we will live tomorrow, whether we can live through the threat to life and civilization.

I think that when one comes to talk about tomorrow, there is a sense that the book of the past is closed, that one has a fresh page to write on. There is a sense of delight and hope that one is free of all that has happened, and able to make it better. I know that in 1945, after the war was over, not only in Europe, but in the Pacific, some of us met to talk about the future. We had a great sense that the terrors that had marked the preceding decade, that dreadful war, Hitler, blood and death, were past, that we were looking out at a future which was ours to hope. We even thought about atoms for peace; we thought about them much more gaily than is possible now. [. . .]

One has these moods as after a severe illness, a long convalescence, after being in battle, and recovering from wounds, of looking to the future, untrammelled by the past. What Baudelaire said of death, we think of tomorrow: "qui refait le lit aux gens pauvres et nus." But it is not really so. Our morality, our politics, our law, our art, and even our revolutionary science are rooted in

157

the past and the present; they do not determine, but they condition, what can happen.

Indeed, the pages are not blank that we have to write on, that we have to read. To me, it is not clear, not in the least, whether we shall live tomorrow. I think it only reasonable to welcome the fact that we have lived for twenty years without great war and to welcome the increasingly wide and deep recognition of the danger that lies in war. This recognition has been made formally manifest in some arrangements between the Soviet Union, the United States, and other powers—the test ban, a few other things, and even in the very existence of the Disarmament Commission which continues to meet in Geneva. We can very much welcome the substantive changes in military policy of the great powers, which reject any simplistic view, such as the rather dangerous one of massive retaliation. But the dangers continue varied and acute, and both they and the measures to reduce them have much to do with how we will live in the future.

There are many other things that we cannot erase from the pages when we look toward the future. Now is a time when we are aware, I think more acutely than for a long time, perhaps ever, of social injustice in all its forms. We are aware of the fact that freedom, even by the modest definition which we accept, is by no means universal in the world; there are acute injustices within countries. I think of the United States, where the signs of this, disturbing, are nevertheless also necessary and encouraging. Now is a time when there is a new sense of injustice, again welcome, but full of danger, between nations. I think of the fact that it is just a quarter of a century ago that the planes and tanks moved eastward from Germany to Poland, ushering in a disaster which is a disgrace to our civilization and our heritage, from which we have still not recovered. In trying to help, or in watching other people try to help themselves, the cultures with which they have lived, fragile, unstable cultures, are being destroyed, and the great strong western culture of which we are the representatives often has very shallow roots to replace what has been lost.

Other people would make different lists. I think of the fact that we have not come at all to grips with the humanization of work in a technical and automated society. We do not know yet what this will mean for us. I think of the fact that all about us there are, there may always be, signs of savagery and hatred, hardly compatible with what we profess and what we hope. I think of Proust's phrase, that "indifference to the suffering of others . . . is the one true permanent form of evil." [. . .]

There is complementary danger in foreseeing the future, and trying to foretell it. We are often convinced that yesterday's surprises will determine what happens tomorrow. We are often convinced that if we only make it go more so as it has gone in the last year or decade, we will see the future. But for good or ill, tomorrow is novelty. It is the novelty of chance, things that come together in a way one cannot predict. Even more, it is the novelty of human imagination, love and design, dedication and purpose. What makes tomorrow tomorrow is that it cannot be foretold today; it is not implied by today.

So, doing our best and drawing on history and political economy, on what we know of psychology, we shall probably miss the point. We will give a varied, a plural, an incoherent account, I hope illuminating, I am sure honest, assessments of what we see today, what we see in the immediate past, that will limit, will help to determine—but will not determine—what will happen tomorrow.

My title is an ambiguous one. The word "common," "le mot 'commun'" means what we share, what is for all men or for many men. It also means "vulgar," "vulgaire." This is a comment on the time. One ought to have a word which means "humanitas" and does not mean "vulgar." [. . .]

I have in mind in using the word "common," "commun," that it refers to what is general, accessible common knowledge, neither secret, nor restricted to an elite; knowledge that is not private to a man, to a family, a profession, or to any of the many communities which span this earth.

[. . .] We should be aware that cultures differ in no way more radically than in where they decide that something is public business, the business of the village or of the community, and whether it is the individual's. Even the most elementary physical aspects of life are assigned public or private, solitary or familiar roles, in a way which varies from place to place, perhaps described by anthropologists, but not easily understood. It is so extreme that students of traditional Japan have found in the absence—almost total absence—of private life, in the totally public character of human existence in Japan, the origin of the custom of suicide for

159

reasons of honor. It was an isolated community, with no place to go, no place to hide, no place to flee, no place to be private, no escape, besides death. In general, I think it is true that the more isolated and the more primitive communities have had a great common culture and reserved less for private life than we are used to or can easily imagine.

What should be private? What should be public? What should be kept secret? [...]

The fact that knowledge is open, as it normally is in scientific things, for instance, or as it normally is in art, the fact that it is public, printed, does not really mean that it is shared. It is very much a characteristic of our time, of how vast the world is, of how fast it changes, of its important and beautiful powers of specialization and language, that all channels are overloaded, and all spirits struggle for light, clarity, and depth against the flood of the information, the creation, the discovery, and the invention that is available. This struggle one must not lose. I have often, among many others, spoken of this, and especially as it affects the life of the arts and the sciences. We all know that something can be and is being done about it by education, above all by the sense of lifelong education and learning, by teaching in all its manifold forms, by the catalysis of friendship and love. We all know that these help to keep our spirits at the same time open, so that we are not shut off from what is open knowledge, and still to keep them honest and intimate and very deep. [...]

Science at its best, and really inherently, is both intimate and open. It has a remarkable sunniness, a warm quality. It has, of course, its rivalries, its agonies, its struggles, its partisans, but not really, I believe, any barriers of any apparatus of secrecy. There is nothing in it, nothing even marginally relevant to it, involving hate, or anything dark or evil. It has no way to suppress and no way to exploit. Of course, it has its sorrows, but so does all life. I remember what Wolfgang Pauli said when he was very young; he was a very early genius. When he was studying in Copenhagen, Niels Bohr's wife asked: "Pauli, Pauli, why are you so unhappy?" And he looked at her angrily, and said: "Why shouldn't I be unhappy? I can't understand the anomalous Zeeman effect." In this sense, science is not always happy, but compared to all other life, I think it is.

160

Of this field, I think one can always prophesy in the sense that we will know more, and usually that one can identify the growing tips of knowledge, that one will find as many new questions as answers, that one will be amazed, and when things are really good, shocked and filled with wonder and rather shaken. [. . .]

Yet once it comes to the practical powers that come from science and its techniques, there are quite other notes sounded. Here prophecy is relevant and here, I think, we ought to spend our time seeing what we can make of the future. Here there is also a sense of responsibility for the future, for choice. Prophecy and responsibility are related by complementary modes; the one says: "this will happen; it must"; the other says: "this should happen; and I will work for it." [. . .]

[The ideal of an open world] is very old, but was newly sharpened and given new meaning some twenty years ago, when the first atomic bombs were being developed. At that time, in mid-war, I think there were two reasons that one went into this. The first was Hitler. I think it was a good reason. The second was also, I think, a good reason, and that is that this development would change the world, the world of Ypres and Verdun, of Guernica and Warsaw. One hoped that it would be changed for the better. Later, in our country, as it was clear that Hitler would be defeated, there were I think two views of the future, probably shaded and mingled. One was that these instruments, if they worked, would help keep the peace. The other was that they would lead, that they should lead, to an international scientific and technical collaboration, which would serve to control their dangers, which would help to transcend national powers to prepare and to make war, national powers to isolate people and engulf them in lies, as Hitler had done, and Stalin. There were many who thought such thoughts and hoped such hopes. One man saw somewhat more vividly how deep the changes would have to be and struggled to bring them about. He was Niels Bohr. Bohr in 1939, after years of work on the structure of the atom, came to the conclusion that making bombs from fission was a pretty tough job. When, in 1943, he escaped from Denmark to England, he learned from Chadwick and Sir John Anderson about the bomb development operations taking place in the United States. This made a very great impression on Bohr, because unknown to him, great industries had been developed to make the bombs; he was persuaded that they would succeed. He had a great vision then. He understood that these weapons, if they could be made, could be made bigger, and could be made plentifully, and that their secret accumulation, their accumulation in closed societies, societies that had no converse with one another, would pose a threat to the security of the world and all its peoples which would be intolerable. He thought an atomic arms race, and he was sure that the Soviet Union would be the major protagonist, would be intolerable. He did not think that we could live through it, as we have for two decades. He did not want us to, and in that he was certainly right. But he understood immediately that this could not be prevented just by signing treaties saying: "We won't make such things," or even by having commissions of experts meet from time to time to control the observance of these treaties. He thought that any real security would have to lie in a deep change in the world, a change which first of all implied that nations would not try to destroy each other, or to conquer one another, and which then implied that all nations would be open to each other, insofar as what went on in them would concern the security and the welfare of men. Nothing of import to that security and welfare should be kept secret.

Bohr knew, of course, that there are moments in a man's life when things have to be private and secret. Essentially he knew from his own experience, as we all do, that when we approach a problem, we usually make mistakes. The only way to try out a course of action, the only way to see whether an interpretation of nature is right, is to talk about it, and not to be held responsible for advocating something that one has proposed. This is essential in government; this is essential in any group that has respon- sibility. Of course it has a deeper and richer meaning in the life of the individual man.

Bohr thus hoped that the allies, the United States, the United Kingdom, would approach the Soviet Union—he regarded this as the vital step—to see whether they could not be interested in attempting a new form of international and national life in which these weapons would not be made, in which things would be open, and in which we would renounce the use of the weapons, the weapons themselves, and offer to help with all the technical enterprises which might open up from this and other branches of science, working in cooperation in a world in which there would be no secrets, where secrets would indeed be illegal. Bohr talked about this with Roosevelt, apparently successfully, and with Churchill, obviously not successfully, and with many of their advisors. [. . .] As it turned out, Roosevelt and Churchill— Roosevelt by then no longer well—partly misunderstood, partly distrusted and indeed thoroughly rejected this suggestion. The efforts that were made to revive it after Roosevelt's death also came to nothing whatever.

It is hard, looking back, to take a very sanguine view of what Stalin would have thought of it. Bohr knew enough of dictatorships to assess the great part that secrecy had played, and would have to play, in their maintenance. He knew that a contempt for truth really played a very large part in the whole nature of the Soviet regime, and its stability. He still would have liked to work for an open world.

Today, twenty years have passed, and it seems to me—I speak with great timidity of this—that although the political grounds within the Soviet Union for being troubled about an open world have lessened, they are still quite far from being trivial. I think that probably the military grounds are gradually being eroded by technical intelligence. Of course, today less than ever is this a problem for just two countries alone.

We may live with the arms race and deterrence, with its very great and rather incalculable dangers, dangers of error, of accident, of folly. Perhaps the time to avert it has passed, or never was; other great catastrophes or changes will have to come. If we should ever move, and I think we should not at all give up hope, to a less dangerous world, it will indeed have to be an open one. From the fact that there are no bombs, it will follow that there will also be no great national wars, and no hopes of conquest or dominion by force or stealth, and perhaps not even any sanctioned and even hallowed hatreds of one nation against another. There are, even today I think, some limited, very slight signs; I think not so much of the formal agreements, limited and tentative, symbolic though they are, and welcome though they are, as of the obvious intention of the armed powers to reduce the probability that they will find themselves engaged in this great new slaughter.

The other side is that there have been some changes, obvious when you think about it, but not much described, in the status of the privacy of an individual. It has always been an essential part of our lives; and increasingly, as the world has grown, and grown larger and faster and louder and windier, we have treasured it. Privacy of the person, of the family, of the group of colleagues, of the profession, anything that gave intimacy, deep meaning, friendship, simplicity to human intercourse. We need it to make our errors without disaster; we need it even to find out what we mean, to talk with one another freely, and for the health of the human spirit. We have always needed it very deeply.

In conflict, privacy is always invaded. There is the "police dossier," the private detective, the informer, the practices of seizure and search and sometimes of torture, whether for power, for knowledge, for battle, for money, for love. The most benign techniques can be used to invade privacy. Let me give one example: for medicine, in order to find out what is happening in a man, very small capsules which will for instance measure acidity and transmit by radio what they have found, have been developed. These are very fine for finding out where a man is, and they are very fine for finding out other things about him; you can conceal them, so he swallows them without knowing. Another example: the tracers, the radioactive products which are the first early major scientific and technical by-products of large-scale atomic energy, are very readily available and are frequently used to follow people when one wants to know where they have gone, and does not want to be seen following them. And so it goes. The circuitry that has been developed for rocketry, micro-miniaturization, information retrieval, have led to the possibilities of surveillance which do not reveal themselves to the victim, which make it possible to follow, to see, to hear, and often to interpret what is going on. These of course are primarily things that one would expect to find in intelligence and counterintelligence agenices. But they are for sale, and cheaply and abundantly and more and more effectively. I do not mean to pretend that there are not countermeasures, or that all devices work all the time; but generally speaking, privacy is not something that you can have, unless you have rather a good technical apparatus to protect you, or unless you are just not interesting, and nobody wants to find out about you. There are also medical applications; there are drugs which make you say what you would not say if you had not taken them; there are effective uses of subliminal perception which make you behave in ways in which you would not if you were aware of what was being done to you. This is a very rapidly growing affair, and so far in our country rather completely uncontrolled.

There are of course, and with us this is very important, legal restrictions on what you can do with evidence like this in a court of law. You cannot introduce evidence obtained in many of these ways as valid evidence in legal proceedings. There are constitutional guarantees in most countries of the western world against unwarranted search or seizure or detention. But the guarantees are of course far from universal; and they are not relevant to the private feuds and struggles of a pluralistic society. I do not know how widespread and how subject to control these possibilities of invasion of privacy will turn out to be; perhaps rather little. Of course, no one may be sure that he is without privacy. But if a man is interesting, if he has an enemy or is a trouble, I think he will not be very sure, unless he is so well equipped, technically, that his actions, his word, his appearance, and what one may conclude from all of this will be unknown to others. This inspires in most people a certain revulsion.

A decade ago, the Atomic Energy Commission in my country did investigate and hold hearings on my trustworthiness; when the proceedings were published, many said that my life had become an open book. That was not really true. Most of what meant most to me never appeared in those hearings. Perhaps much was not known; certainly much was not relevant. I did have occasion then to think of what it might have been like to be an open book. I have come to the conclusion that if in fact privacy is an accidental blessing, and can be taken from you, if it is worth anyone's trouble, for a few dollars, and a few hours, it may still not be such a bad way to live. We most of all should try to be experts in the worst about ourselves; we should not be astonished to find some evil there, that we find so very readily abroad and in all others. We should not, as Rousseau tried to, comfort ourselves that it is the responsibility and the fault of others, that we are just naturally good; nor should we let Calvin persuade us that despite our obvious duty we are without any power, however small and limited, to deal with what we find of evil in ourselves. In this knowledge, of ourselves, of our profession, of our country—our often beloved country—of our civilization itself, there is scope for what we most need: self knowledge, courage, humor, and some charity. These are the great gifts that our tradition makes to us, to prepare us for how to live tomorrow.

At the San Diego Zoo around 1938. Second
from the right Robert Oppenheimer; on his
right Luis Alvarez who worked with
Oppenheimer on the bomb project and later
was to testify against him. On the very left:
Robert Serber. (OMC)

14

TO LIVE WITH OURSELVES

1965

[...] I USE THE WORD "SCIENCE" to mean really all those areas of human knowledge, still a small part of human life, all those areas of human knowledge where we can tell each other what we have done and what we have found. This knowledge is historical, sociological, economic, mathematical, anthropological, astronomical, among many other forms. It is in just such fields that you have been working. I know that the word "humanities" can be used to talk about the whole range of expression that men give to their experience. I think the analogies that bind archaeology to astronomy are not very much more remote than those that bind anthropology to astronomy; I speak to you in this sense as fellow scientists.

You know that great precocity in science is quite rare; there is no Mozart; and this is not entirely astonishing, because you have to get a lot of partly wrong ideas in your head before you can find out that they are wrong, and that is the beginning of scientific inquiry. It is true that Galileo, though many of the stories about him are not true, was really eighteen years old when, using his pulse as a clock, he established the synchronism of the pendulum, saw that the time of the oscillations did not depend on the amplitude of the oscillations. This was with a lamp that hung in the cathedral at Pisa; then he went home and tried to build somewhat less bizarre clocks of his own, with which he later learned to time the water clocks that he used in establishing the law of falling bodies.

Galileo's life shows that a man is much more than a scientist, even when he is wholly and devotedly a scientist. His life was not free of sorrows. The later days, when he lived in some fear and suffered the great indignity of his abjuration, and alienation from many of his friends, were certainly sad years. But looking over his life, rereading what he wrote, even very close to the end, there is a quality which I believe to be true of science in the making: a great sunniness in the act of finding out new things, things that he had not quite expected, things that reflected deeply on beliefs long held, and that pointed both in hope and in mystery to the future.

[. . .] Whatever trouble life holds for you, that part of your lives which you spend finding out about things, things that you can tell others about, and that you can learn from them, that part will be essentially a gay, a sunny, a happy life, not untouched by rivalry, maybe not even untouched by an occasional regret that somebody else thought of something that you should have thought of first, but on the whole, one of those nobler parts of the human experience. This makes it true that the life of the scientist is, along with the life of the poet, soldier, prophet, and artist, deeply relevant to man's understanding of his situation and his view of his destiny.

It was in 1935 that I took off a little time to go with my brother and a friend to a primitive ranch that we have high in the Sangre de Cristo Mountains not far from Santa Fe. That summer we decided that we should have a look at the San Juan Mountains; so we saddled up our horses and rode, in what was a rather naive straight line, to where we were going, and it was not until two nights later, in what was then a rather poor town, Tierra Amarilla near the banks of the Chama, in northwestern New Mexico that I got to open the letter. The letter told me of experiments that Milton White was conducting, and that had just reached a certain stage of clarity. He is now the director of the Pennsylvania-Princeton Accelerator in Princeton; but he was then a graduate student. We had talked to him about the fact that the law of force between two protons should not be exactly Coulomb's law, the inverse square law. This was because by then we understood that around the protons there would be a charged cloud of electrons and positrons. These were small deviations, and the technique that White used would indeed have taken him a long time to find these; only much more recently and by much, much subtler approaches have they been confirmed. But he had found something. He had found that there were enormously more large angle scatterings of protons by the protons in Hydrogen than the electrical forces could explain. This was the first clear demonstration of nuclear forces between protons. That some such thing must exist was clear for a long time because nuclei exist; but the very great strength of the forces, their short range, all were apparent from these first primitive experiments.

This was one of the many times when the question, "how hard is matter?" got a new, fresh answer. At these distances and for these energies, matter was very much harder than had been known before. This was something Newton speculated about, and, talking about what holds the molecules of matter in place, he had spoken of the inconceivable hardness of these forces and the infinite hardness of these forces. This new discovery paralleled in some ways Rutherford's discovery twenty-five years earlier of the existence of the atomic nucleus, and paralleled very much studies that are now in progress, and still not conclusive, to see how hard the nucleons themselves—the ingredients of nuclei, protons and neutrons—are. Do they have something very hard in them? We do not know; we have not seen it and it is probably not there.

Rutherford's discovery was not, in the first instance, his. You may know this story. He had spent two decades unscrambling the radioactive radiations and the chemical properties of the heavy elements that emitted these. In the course of this, he had gotten used to thinking that atoms were not very hard, not at all like Newton's conjectured ones, not like the atom of classical atomists, because one of the radiations, the alpha particles, went through the atoms without doing very much of anything. They would gradually lose their energy, they would be very, very slightly deflected, but they paid no attention to the atoms. They just treated them like mush. But Rutherford was not someone to leave well enough alone: he had a young student called Marsden, who seemed to need a doctorate; thus Rutherford gave him a problem which was not, to Rutherford's mind, very promising. He said, "Why don't you look and see whether alpha particles are ever deflected by really large angles? Do they ever really turn very much away from their path?" Marsden worked on that, and it did not take him very long. The equipment was very simple. He worked with Geiger; and they came back very shortly afterward and said to Rutherford, "The alpha particles are sometimes scattered back at us." Rutherford said at that time, "This is the most incredible thing that's ever happened to me. It is as though you fired a sixteen-inch shell at a piece of tissue paper and it turned and came back at you": so radical was the notion that there was something hard at the center of the atom. Everyone had thought until then that there were electrons in an atom, that are not very hard, and cannot scatter an alpha particle, and that the positive charge was diffuse. This was not true. The positive charge was highly concentrated; the forces, as Rutherford figured out, were not nuclear forces; they were familiar electrical forces. It took Rutherford a year and a half, more or less, to figure out what it was that Marsden was observing, and to check it out by predicting how the scattering should depend on the atomic weight of the material, the angle of scattering, the energy of the alpha particle. In 1911 he came out with it, and that was of course the start of atomic physics.

I remember that four years after the letter from Milton White I was back at Berkeley, and a colleague, Luis Alvarez, came to my room, very excited, and with a good deal of blood showing on his neck and little pieces of paper stuck to it to stop bleeding; and I asked him what was wrong. He said, "I have just been to the barber's, and I read in the paper, uranium comes apart in two pieces when it is bombarded with neutrons!" This had just been discovered in Europe, and talked about by Bohr, who was then at the Princeton Institute. It again looked back into the past, because no one had taken seriously this kind of nuclear reaction, radically different from all those that had been studied. It looked to the future because, although neither Alvarez nor I could then foretell or be confident that there would be practical consequences, it occurred to us that this might have consequences of the gravest interest, gravest importance, and perhaps, I would say, greatest hope for man.

The first time that I had anything like this happen to me is not recorded in my citations or biographies, but I remember it. I was a student at Göttingen, and it was the first time that something came over me which I had not thought about, which nobody else had talked to me about, and which I think, in that form, at that time, nobody knew. This had to do with what happens to a Hydrogen atom in an electric field. It had been discovered long before by Stark that the spectral lines are split and slightly displaced; and the new wave mechanics had given a quantitative and correct account of this; but when I tried to see what were the stationary states of a hydrogen atom in an electric field, states analogous to the ground state of the atom or one of its excited states, I found that there were not any and that puzzled me for a little. It need not have, because if you have a constant electric field, then the potential is represented by a straight line. On one side the potential is low for an electron; on the other high. As for the hydrogen atom, you may think of it schematically as making a deep hole around the position of the proton; the state of the hydrogen atom will have an energy that is matched far out where the potential is low. The reason that there are no stationary states is that the barrier, the high potential mountain which separates the two regions of low potential, is not completely opaque; when you take into account the wave nature of matter, no barrier is ever completely opaque. Thus there is a trickle of probability for the electron to emerge and fly off, and that is why the atom is not stable. This interested me because I thought it would have to do with the conduction of electricity in metals. The first application I made of it was to an effect which is not important, but amusing: you can pull electrons out of metals with an electric field; and I figured out how the current should vary with the electric field. When I came to Cal Tech, I found that Millikan and Lauritsen had discovered this law as an empirical law.

That was not the important thing. The important thing was that people who had been worrying about nuclear physics realized that here was the solution to an old paradox of alpha decay. In alpha decay, a nucleus loses an alpha particle; its weight goes down by four units, its charge by two. If you bombard the daughter nucleus with those alpha particles, nothing happens except that they find an electric field. They almost never get back in. Here you have a situation in which the alpha particle has a repulsive field around the nucleus; obviously something attracts it because it stays in the nucleus for a while, so it may have an almost stationary state, when there is contact with the rest of the nucleus. Here again is a barrier; and the same laws describe the leakage through it. That was done by Gamow and Condon and Gurney. Even that was not very interesting; but it showed that particles could penetrate into nuclei when they did not have energy enough to come over the top, when they simply leaked through the barrier of the electrical potential that repelled them. That led to two things: the first was insight into what makes the stars shine, because this is what goes on in the stars, in the sun and most other stars. It also led people to build accelerators, and get protons that could get close enough to nuclei to leak in, although it took a long while to get them fast enough so they would go over the top and fall in. Here again things that had been puzzling for a long time fell into place and, rather sharply, something new was open for the future, both in astrophysics and in the experimental breaking open of the field of nuclear physics.

There are a number of such examples, even in my life. The positron was more or less predicted, and it was found in 1932. Then we took quite seriously the question of what would electrons or gamma radiation do in cosmic rays, and saw that what would happen is that the light would make electron-positron pairs that would pass through matter. The electrons and positrons would in turn radiate and make gamma rays; starting with a high energy particle, one would very soon have an enormous family or cascade of particles. It was known that such things happened in the cosmic rays; this fit rather well. It had to fit. Yet there are lots of particles in cosmic rays, both positive and negative, that do not do this; they just do not make cascades. Such particles were not then known to exist in nature; so the success of this theory of cascades meant clearly that there were new objects in the cosmic radiation which men had not seen or identified before.

We know today that there are hundreds of such objects, but only a few found in the cosmic rays. The last time that the cosmic rays taught us something new about the structure of matter was about fifteen years ago, when another unexpected discovery occurred. One would see, pointing up, two divergent tracks in a cloud chamber; that was called a lambda, because it looked like one; or one would see a charged particle coming in and then, without any scattering, a different charged particle coming off and nothing else visible. These were obviously pictures of a neutral particle decaying into two charged ones, and of a charge particle decaying into a charged and a neutral one (the one you do not see). These were soon identified. One for instance could be a kind of meson, a kind called pi, and the other a proton. This immediately posed a very serious problem: here is an object, the neutral parent, which changes, though not very rapidly, since it has been travelling quite a while, into a proton and a meson. Why does it take so long? There is plenty of energy for this; both products move fast. There is no potential barrier; the decay is caused by a very weak force, very weak compared to electricity. Then how could you make a lambda, which is not a normal ingredient of matter, does not last very long, could not come in from outer space, how could you make it from the materials through which the cosmic rays pass, from the protons that they themselves are, and the material, air or lead plates or whatever things they traverse. If you could make it only so very slowly you would never see it. If you make it faster, why does it not come apart fast? These questions led closer to some quite striking developments. One learned first that there are hierarchies of forces in this game, some a thousand billion to a million billion times weaker than others. The strong forces that are involved in producing these particles, then called strange, are very much stronger than those that produce the decays. One learns something deeper, which took quite a while to sort out and is not fully sorted out today, and that is that the different order of forces have quite different hierarchies of strength, and quite different rules as to what processes they allow and do not allow. The strong force that produces the lambdas produces another object, often not immediately visible, but often detectable, which has a quality which compensates for the strangeness of the lambda. In the decay process there is no such conservation of this strange quality; that process goes on very slowly.

We are today living in a marvellous and really quite unfinished business of trying to understand the order of these forces, the relations between them, if any; we do see some. But here was another beginning, I think the last subnuclear discovery to be made by cosmic rays, which again showed that our views of matter had been much too simple, and which multiplied questions for the future.

The history of the sciences, and even today the history of the humanities, are full of wonderful stories, many of them legends, pure legends of discovery. You will think immediately of Archimedes and his bath and his word "Eureka," and of Kekulé travelling on the bus in London and thinking of what the benzene ring might look like, something that really became intelligible only after there was a quantum mechanics. I have a few such things in mind in stressing what I know you must have seen, in trying to encourage you to believe that it is the trait of *science in the making*. I think that in all of Galileo's life what most shook him was the discovery of sun spots. It is not the most important thing to us today; it was not for the history of science in its development at all the most important thing. Just as Newton's universal law of gravitation changed entirely the division between the terrestrial and the celestial world, with which scholars had lived for two millennia, so the imperfection, the confusion, the chaos of the actual sun put an end to the idea that there was anything perfect or unmaterial in the heavenly bodies.

When Mendel found (and we know that his numbers were indeed too good, but we attribute that to the benignity of the young monks who helped him) that the distribution of populations in his peas were the same numbers as the coefficients in the binomial expansion, he came upon something which showed an element of discreteness in inheritance, and in all life, which is of course today one of the richest fields of discovery.

In 1905 Einstein, working alone, hardly knowing a physicist, never having had anything much of a conventional education as a physicist, working in the patent office in Bern, that year, discovered two things, among several others. He discovered the meaning of the velocity of light, as the limiting speed at which signals could be sent, and the profound change that this brought as to the nature of measurements, of simultaneity, and the effect of these changes on the description of relative motion. In the same year, by seeing what his other love, thermodynamics, implied for the validity of the law of radiation that Planck had found, he discovered that for that law to be true there must exist quanta of light, with an energy proportional to the frequency, and a momentum inversely proportional to the wavelength, and that their presence showed itself in the fluctuations about this equilibrium in which individual quanta appeared and disappeared and contributed a fluctuation proportional to the square root of their number, just as independent atomic events.

It was in 1911 that Bohr met Rutherford, and found out about the atomic nucleus. He also had one of the revelations, full of paradox and incomplete for another fifteen years, but again, to a certain extent, transcending all that had been thought in the past about what an atom looked like, and opening the way to marvellous developments in the future. Bohr had a look: if the nucleus were really there, the electrons had to be somehow around it; and the nucleus was very small compared to the region they occupy. In classical mechanics and in classical electrodynamics there is no place for such a thing: it does not work; you cannot get reproducible atoms; you cannot get stable atoms; you cannot get anything like a real atom. Bohr saw at once two things: first, that the electrons had one game and the nucleus another, and that the two problems were, in very good approximation, separate, so that you were free to think about these two orders separately. Second, he saw that the atomic problems, the problem of what the electrons were doing, granted that there was a nucleus, could not be understood unless, as Bohr put it, the quantum, the same quantum that Einstein discovered for light quanta, governed the laws of motion of atoms and determined their stationary states. One can see that you cannot define quantities that are the right size to be an atom, or that radiate colors that look like atomic colors, without using Planck's constant to express them.

I have mentioned mostly physics, in which I have spent much of my life. My colleagues at the Institute sometimes tell me their stories. It is only in the last years that a tablet was found, quite by accident, in a schoolhouse in Troizein, which is a copy, its authenticity still debated, but I am partisan, I believe it authentic, a copy that the citizens of Athens had made for the citizens of Troizein. During the fall of Athens and its occupation, Troizein had invited women and children from Athens as refugees; in gratitude, after their liberation, after the war was over, the Athenians had the order of battle that Themistocles drew up before the battle of Salamis recopied for them. This is, perhaps more than any other battle, one that determined that there would be a European culture, a culture based on Mediterranean civilization, a culture of which we are the blessed inheritors.

Again, a few years ago, there were some diggings in the fields of Lavinia, and there came the first hard evidence that the story of the trip from Troy to Rome was true. One thing I remember seeing is a photograph of an inscription from an altar piece. The first three words are in Greek, the next two are hard to read, and the last two are in Latin.

There are discoveries to be made about our past, even about our very recent past, which confirm conjectures, which show that we were all wrong about what we thought, and which open up new avenues of study and new prospects for human life. Whenever this happens, when the discovery has any of the qualities of the great ones, it has to reach back into a solid framework of experience and understanding and a great tradition; it has to mean something. It has not to be just something that could or could not be true, that might or might not be true: it has to imply. Often of course it is a discovery that confirms what people thought. Sometimes our ideas are right; that does happen; but more often and in the important cases these discoveries upset what was thought. They bring a sense of shock, and when they are very great, or when we who make them think they are rather great (that is not the same thing), they bring a sense of terror, a sense of terror because of the destitution in being cast loose in a new unknown, a sense of terror from knowing what is ahead. In the last years of his life Galileo wrote about this: He said "With our infinities . . . we are adrift at sea, and who knows whether, no matter how long we may dispute, we shall reach firm land again." John Donne wrote about it somewhat earlier, " 'Tis all in pieces, all coherence gone."

I have heard from some of the great men of our time that when they found something startling, they knew it was good, because they were afraid. Otherwise there is nothing much to say; there are so many different ways it happens; a new instrument may lead to a discovery, a new accident, a new argument, a new finding, a measurement just a little more careful than was made before; looking at a problem in a fresh way. What all this does is of course to add to what we know and so add to our responsibility when we come to use that knowledge.

What it is which marks this part of human life from, let us say, that of the poet, is that science is the business of learning not to make the same mistake again. We go on making mistakes; the art is to make the most fruitful mistakes we can, and hope that we find out soon, and finally do find out. But you do not make the same mistake again, because you tell what you have done; people try it out; they see if they find what you said; and when this has happened then that will be firm. There may be other things that one did not think about, that have still to be done. Yet in this sense there is an improvement of man, in the sense that he does not make the same mistakes again, in what he knows of the world, and of his role as a knower. This of course helps to make us more aware, both of how perpetually faced with problems, with ignorance, with darkness, with mystery, we will be, and what it is, really to live in the world of nature, and with ourselves as a part of it. It helps us to know, though it does not determine for us, what we may hope to do, and what we cannot do. It even helps us to live a little with one another and, not always, but at best, it helps us to live with ourselves. [. . .]

At a panel discussion in 1961. (Photograph by P. Karas, courtesy of MIT Museum)

15

PHYSICS AND MAN'S UNDERSTANDING
For the Smithsonian Institution Bicentennial

1965

WE ARE CELEBRATING A BIRTHDAY, honoring the foresight of a man and the success of a great institution. This makes it fitting that we leave to one side the common plaints of our time: that physics is corrupted by money; microbiology and mathematics by pride, not unrelated to achievement; astrophysics and geophysics by access to novel and powerful instruments of exploration; the arts by alienation; and all by our lack of virtue. What truth there is, and there is some, to these anxieties is not for us today. We could begin with Joseph Henry, the first Secretary of this Institution, [. . . who said,] "Knowledge should not be viewed as existing in isolated parts but as a whole, each portion of which throws light on the other . . . the tendency of all is to improve the human mind . . . for they all contribute to sweeten, to adorn, and to embellish life." When we think back on the prolonged and troubled debates with which the Congress moved toward accepting Smithson's bequest, establishing this institution, we can only be moved to celebrate the extent to which it has managed to preserve and enlarge, not perhaps the unity, but the harmony between the sciences, between the arts and sciences, between nature and man, and between knowledge and practice, whose conflicts so troubled the Congress for almost two decades.

Physics has played a part in the history of the Smithsonian, as indeed it has in the history of the last five centuries. Closely related to astronomy, to mathematics, and to philosophy in its earlier years, it now has intimate relations also with all branches of science, and plays an increasingly explicit, conscious and visible role in the changing conditions of man's life. [. . .] If physics has had these extended relations with science and practice, it has still maintained a kind of central heart of its own. This is because it seeks the ideas which inform the order of nature, and of what we know of nature. Countless phenomena which, from the point of view of physics, appear calculable and explicable, but not central or essential, turn out to be pivots of our understanding in other sciences. No *a priori* study of physics would have been likely to explain the accidents that make the synthesis of carbon in the stars possible. Yet that has made a difference of some importance to man. Most of the miraculous findings of microbiology were not invented, and would not have been invented, by physicists, though they have played an appropriate part in helping to provide the instruments and the language for their discovery. For every science, much is accident; for every science sees its ideas and order with a sharpness and depth that comes from choice, from exclusion, from its special eyes.

[. . .] What has happened in this century in physics rivals in its technical and intellectual imaginativeness and profundity what has happened at any time in human history. Its effects on the way we live are even more immediate and manifest than was the use of the magnet for navigation, or of electricity for communication and power. It has now, however, led to so great a change in man's views, of his place in the world, his function, his nature, and his destiny.

The years from the thirteenth century to the seventeenth saw the gradual acceptance of a material world no longer centered on man, or on his habitat, the gradual acceptance of an order in the heavens that could be described and comprehended, that sharply limited and circumscribed, though of course did not eliminate the role of God, or indeed of accident. We should ask ourselves why the views of Copernicus, the discoveries of Galileo, the understanding and syntheses of Newton, should so greatly have resonated through European society, so greatly altered the words with which men spoke of themselves and their destiny. Nothing like that happened with Einstein's theory of relativity, which tells us the meaning of the velocity of light—or of quantum theory, which tells us of the meaning of the quantum.

[. . .]There is an analogy, long known to physicists, between the special theory of relativity and the quantum theory. Each is built about a constant of nature and has something to say about how the constant, in determining the laws of nature, restricts or enlarges our ability to learn about nature. [...] Einstein's first theory of relativity made clear an unexpected meaning of a constant of nature long ago determined by the astronomers, the velocity with which light propagates in empty space. It was Maxwell who showed that this constant was the same as that relating fundamental electric and magnetic units, and explained why this should be so, by showing that light is an electromagnetic wave. Einstein's role was to recognize that because of the universal validity of Maxwell's equation, and the independence of the velocity of light from the velocity of the source emitting it, this velocity must, itself, take on the role of what in earlier times was regarded as an infinite one, one which could not be surpassed. The corresponding limitations, the absence of absolute judgments of simultaneity at distant points, struck rather deep at all views of space and time ever held before. At the same time, they liberated physics to form new and consistent descriptions of nature, and by altering and refining Newtonian mechanics, to anticipate new interconnections of the most fundamental theoretical and practical import.

In some ways even more remarkable was the interpretation of Planck's constant, the quantum, that emerged from the development of the quantum theory of the atom, the work this time of many men, initiated in part by Einstein, in part by Bohr, and brought to an essential clarity by Bohr and his Copenhagen school. Here, again, physics was given a great liberation, the ability to understand the stability of atoms, the atomicity of matter, the regularities of chemistry, the atomic and molecular requirements for life, most of what physicists and chemists had known until the turn of the century. Here, again, it was discovered that the role of the quantum in the order of nature limited the traditional concepts of what we could learn about nature by experience. The quantum defines the irreducible roughness in the relations between a system being studied and the physical means—light, or beams of particles, or a gravitational field, for instance—that are used to study it. Because of this, there is an atomicity not only to the atoms and molecules, but to the traffic between them and the physical instrument of the laboratory; and, because of this, a complementary relation of mutual incompatibility between different sorts of observations of an atomic system.

From this follow all the well-known features: the ineluctable element of chance in atomic physics based, not on our laziness, but on the laws of physics; the end of the Newtonian paradigm of the certain predictions of the future from the knowledge of the present; the element of choice in the approach to atomic observation. Yet perhaps the most important lesson is that objective—and massively and beautifully successful—science could be based on a situation in which many of the traditional features of objectivity were absent, and which taught us that for scientific progress and understanding, objectivity is more closely related to our ability to describe to one another what we have done and found, to verify or refute, than to its ontological foundation.

As for particle physics, it is an unfinished story. What we are sure of today may not yet be ready to make its contribution to the common culture. Just from the requirement that in these new domains the general principles embodied in an understanding of the quantum and the velocity of light should still apply, it follows, as has been known for more than three decades, that atoms, or particles, or the ingredients of atoms, could not themselves, as all philosophical atomists had thought, be the permanent, unchanging elements of nature. They are created, destroyed, transmuted, but do not remain unaltered. What do remain enduring are certain abstract attributes of particles, of which the electric charge is the most familiar, and of which two other examples are known: the number of proton-like particles minus the number of their antiparticles, and the same number for electron-like particles. As for several other abstract quantities—such as strangeness or hyperchange, and isotopic spin—that do change but remarkably slowly, we are not ready to tell philosophers what we have made of them. [...] We have a rather unexpected alteration of the ancient atomist's answer to the problem of permanence and change. What lies ahead, we do not know. In the tumult of discovery and conjecture I have great hope; but whether we will be led, as has been so long speculated, to some further limits on what we can say about events in space and time on the scale of the very small, or whether the true shock will be far more shocking, I have an open mind.

[. . .]The discoveries of this century, past and still to be made, find their way into our schools and become part of the language and the insight of new generations, and provide new attitudes and new analogies in looking at problems outside of physics, outside of science, as has already so largely happened with classical mechanics, and with electricity. But it is clear that these discoveries, which were not easy to make, and which, to the professionals involved, brought a sense of terror as great as that which touched Newton, have clearly not changed our philosophy, either in the formal sense or in the homely one. [. . .]

I have sometimes asked myself when a discovery in science would have a large effect on beliefs which are not, and may perhaps never be, a part of science. It has seemed clear that unless the discoveries could be made intelligible they would hardly revolutionize human attitudes. It has also seemed likely that unless they seemed relevant to some movement of the human spirit characteristic of the day, they would hardly move the human heart or deflect the philosopher's pen. I now think that it can be put more simply. These syntheses, these new discoveries which liberated physics, have all rested on the correction of some common view which was, in fact, demonstrably in error; they have all rested on a view which could not be reconciled with the experience of physics. The shock of discovering this error, and the glory of being free of it, have meant much to the practitioners. Five centuries ago the errors that physics and astronomy and mathematics were beginning to reveal were errors common to the thought, the doctrine, the very form and hope of European culture. When they were revealed, the thought of Europe was altered. The errors that relativity and quantum theory have corrected were physicists' errors, shared a little, of course, by our colleagues in related subjects.

A recent vivid example is the discovery of the non-conservation of parity. The error which this corrected was limited to a very small part of mankind. There is a still more recent example, the non-conservation of combined parity, more limited still in the number of us who could be shocked by it, not yet understood, but with hopeful, though still unpublished and unverified, indications of its possible deeper meaning.

Thus I think it is true that only at the beginnings of a science, or only in a society in which an awareness of the problems of science is extraordinarily widespread, can its discoveries start great waves of change in human culture. Just possibly if, in years ahead, other examples, other forms, other sites of life should be discovered, we would have a valid analogy to the great shock of the last century, when the anthropologists showed us the unimagined variety of human institutions. Although the nineteenth-century discoveries in biology had gone far to relate man to other forms of life, although anthropologists had revealed the unanticipated diversity of beliefs, values, and practices in different cultures, and the lack of universality of the ideals by which our own society had been nourished, although the psychologists had brought some supplement to the great religions in revealing again the universal traits of evil in all men, in fact these discoveries were to deepen and not to erode the sense of a universal human community.

If the impact of the developments in physics in this century on the general understanding of man has been restricted, their practical consequences, along with those of all the natural and mathematical sciences, have been unrivaled in their sharpness and immediacy. I should like to mention one instance, in which, largely by accidents of history, the part of physics has been important; that is, the new weaponry, the new situation of the nations and of war. It is still not clear in what way, or even whether, these developments will turn out to be important for human history. I should think it likely that they would be. These developments, and problems that they raise, cannot be lived out in isolation from all others which characterize our time, but they can be talked about in a certain isolation.

It is [now] twenty years ago that men generally learned of the new weapons of a new order of destructiveness. At that time we knew and told our government, as no doubt experts in other countries knew and told theirs, that the bombs that cruelly, yet decisively, ended the Second World War were, from a technical point of view, very much a beginning, not an end. We thought of some ideas about using deuterium and ordinary uranium to increase their power a thousandfold; we thought of the probable appropriateness of delivering such objects by rocket. We did not know too much about it; but within a decade, rather much had been learned.

When I think back to the summer and the autumn of 1945, I remember a number of views of the future which were formulated in this country, and, despite preoccupation with recovery from the terrible war, no doubt abroad. The simplest, and the only one which has been decisively refuted, was that these weapons would remain a monopoly, and thus either play very little part, or put to the test only the restraint, compassion, and fortitude of our own people and government. This was not my colleagues' view of course, nor mine; but for a time, at least, it was that of many, including some of the very highest officers of our government.

Others pointed to the long history of warfare, and talked of a defense against atomic bombs. In no meaningful sense has this characterized any period of the last two decades. As long as the armaments race continues, we will have to ask and reask whether adequate new defenses may be possible. They have not been. Thus, we have lived these years with a complementary and opposed dependence on preemption and deterrence.

Others, looking to past history, trying to look to the future, saw only the certain eventuality of apocalyptic war, postponed in all likelihood by the efforts of statesmanship until it was quite total. This is one forecast that history will never totally disprove. And still others, looking to the past with their eyes, and trying to penetrate the future, held that such self-defeating weapons would be put to one side, leaving the nations to war on one another with more limited means. [. . .]

Yet there were quite other thoughts. Colonel Stimson wrote of *the necessary government of the whole* and Mr. Grenville Clark, then as now, tried to accommodate the needs of world order with the freedom, the diversity and the self-interest of the world's peoples. Einstein said simply that world government was the only answer. To the Acting Secretary of State, the more importunate appeals led him to suggest that it was not always helpful to replace a difficult problem by an insoluble one.

Most of us recognized how central the relations with the Soviet Union would be, and, very soon, how ominous their course. Most of us recognized that with any *government of the whole* capable of serving as a vehicle for common aspirations, for expressing and advancing common interests, the extraordinary diversity of the nations and regions and peoples of the world would present hard problems. There were rich people, and there were very poor people; in any common society these inequalities would more and more become inequities, and the inequities more and more the source of grievance and of guilt. Even in that world which had long lived with the European heritage with a deep—though changing—Christian sensibility, differences of history, differences of political practice, conflicting assessments of the value and meaning of freedom, made talk of the world's community of interest rather a falsetto clarion. We did not know then, but we should have, that in vast parts of the world, in Asia and in Africa, the first, the most powerful, and the most spectacular of Europe's legacy would be the lure of technology, the pleasure of privilege, and the delights of an often synthetic nationalism. We knew that the rich could not, if they would, and perhaps would not, quickly reverse the inequities in conditions of life among peoples. We knew that for the world's future the variety of historical experience, the differences of tradition, of culture, of language and the arts, should be protected and preserved. This left very little of the idea of government of the whole, but it did leave something.

In June of 1945, before the first bomb, Arthur Compton, Fermi, Lawrence, and I wrote, in answer to questions put to us by Colonel Stimson, the Secretary of War: "To accomplish these ends, [the rapid and in human life the least costly of the war, and the preservation of the future peace of the world], we recommend that before the weapons are used not only Britain, but also Russia, France and China be advised that we would welcome suggestions as to how we can cooperate in making this development contribute to improved international relations." These views were endorsed by the Secretary of War's Interim Committee on Atomic Energy, though the Committee, of course, paid little attention at that moment to France, and to China. In fact no meaningful communication was made at all, no attempt to enlist our then allies in a common responsibility and a common concern. That would have been a moment to begin to worry about what is now called "nuclear proliferation," for we and our then allies are the five powers that today have a known nuclear military program. I think that we will not be very successful in discouraging other powers from this course unless we show, by our own example and conviction, that we regard nuclear armaments as a transitory, dangerous, and degrading phase of the world's history, that before other nations could have competing armament, there is a good chance that armament will have become archaic.

In writing as we did in 1945, and then, of course, very much more later, we were not unaware of the diversity of condition, interest, philosophy, and political institutions even in the great powers of the world, and certainly in the world at large. We did know one thing from our experience before and even during the war; we knew something of the universality of the practice, language discourse, and ethos of science. Los Alamos and other wartime laboratories, were indeed international institutions. For years before the end of the war, those responsible for the organization of the scientific effort in this country—Vannevar Bush, and James Conant, and many others—had been speaking of the hope of an international control of the new weapons, and a cooperative exploitation of the new sciences. Similar views were widely held in Britain. Most of all, Niels Bohr explored these possibilities in depth, recognizing that any such cooperation and any such control would have to rest on open access in all countries, and recognizing that this was the best guarantee against the self-delusion and the cultural and political and human abuses of societies that seal themselves off from their fellow men.

The years since the war have brought many examples of effective and fruitful international collaboration, in technology, in political economy, and above all in the sciences. My own field just in the last years has been enriched by contributions of the greatest value from physicists whose countries a century ago were quite closed to the scientific tradition in Europe: Korea, Japan, China, Indochina, to name a few. We need to be grateful for the strength and beauty of this tradition, and to tremble as well as take heart in its power. These same years have also shown how modest, how fitful and inconstant, how easily overwhelmed has been the effect of these international communities on the nations and the governments.

If I recall at this time some notions of two decades ago, it is clearly because I believe them essential to our present and our future. For I see it as a crucial question of our time whether, in a world destined at best slowly to relieve the inequities of rich and poor, the exploitation of military technology, of national pride, of privilege will be met by the growth of a community of interest and understanding. In the discouragements of the day, good example must come to be our firmest ground for hope.

Signed passport photo, taken around 1958. (OMC)

16

A TIME IN NEED

1966

[...] SURELY BY NOW the sciences are for us one common enterprise of communities of men who know one another, work with one another, and are interrelated by a network of bonds of indebtedness stretching between the continents and are beyond disentangling. For a long time now the triumphs have been common, the frustrations shared, and the new problems of growth, specialization, change, novelty, are the shared problems of a common undertaking in which we turn to one another and learn from one another's problems and mistakes.

It is not quite so gay if we talk of culture, or of that domain of action which in the old sense, the sense of Aristotle, is politics. I doubt whether, at this writing, there is among us any wide conviction that France, or the United States, Germany, or England, now has a government even remotely competent to the problems of the time, or, in fact, has available for those problems the human resources, the insight, the wisdom and skill, and the underlying stoic confidence for which they call. [...] This loss of confidence at home and in much of the world is shared by all of us.

Some two months ago a group of distinguished French scientists, most but not all from Paris, addressed a message of condolence and sympathy to some of their American colleagues. They explained that they understood what it was like to live in a country that was conducting the war today in Viet Nam; had they not been through some of the experiences in Viet Nam more than a dozen years ago, and again, far more sharply and with more reason in closer analogy, in the long agony of the war in Algeria? The analogy was overstated and the vast differences unnoticed, [. . .] but the sense of what it is like to live with a government held wrong in a mortal matter is a new and desperate bond between European peoples and those of the United States.

It is not only in the sciences that we have by now so nearly a common history, that we in the United States have for so long owed so much to Europe and its great men, and in our own way have come to play a beneficial part. We are in debt for the two millennia during which Europe preserved and cherished the Christian tradition and the Christian sensibility; for the millennia and then the centuries in which love of freedom and the nurture of freedom flourished; for the adventures, the practical intelligence, the inventions, and their growing and deliberate cultivation that distinguish Europe's culture from the great societies of the East. The spread of knowledge, of responsibility, of acceptance of change, all these are part of the American inheritance. They have been often reshaped, refined, restored, made robust and rather more universal in the American experience. We have given each other capital, and that most precious capital, men; we have given each other technique and knowledge and example, and the forms of art.

We need to help one another maintain a large and common insight and wisdom. We need to help one another avoid the errors that can hang shame about us for our countries' misdeeds. We need to strengthen one another in our human vision of our destiny. This is an extraordinary time, unparalleled, with dangers never before known, with a world in need, the more desperately and urgently felt for our existence and our intrusions. This is a time in need, not of delusion but of hope, when we, not least in Europe and America, who have taught hope so willingly and widely, must bear true witness.

ACKNOWLEDGMENTS
AND BIBLIOGRAPHY

Photographs courtesy of Oppenheimer Memorial Committee (OMC) or Los Alamos National Laboratory archives (LAL), if not otherwise indicated.

Travelling to a Land We Cannot See
Reprinted by permission of *Foreign Affairs*, January 1948 (original title: "International Control of Atomic Energy"); © 1948 by the Council on Foreign Relations, Inc.

The Open Mind
From "The Open Mind", © 1955 by J. Robert Oppenheimer; reprinted by permission of SIMON & SCHUSTER, Inc.

Science in Being . . .
Speech given at Cornell University in the spring of 1949 under the title "The Relation of Research to the Liberal University."

The Consequences of Action
Reprinted with permission of the Association of the Bar of the City of New York, *The Record*, Vol. 6 #3, pp. 106–121, March 1951 (original title: "Contemporary Problems of Atomic Energy").

Uncommon Sense
An Open House
From "Science and the Common Understanding", © 1954 by J. Robert Oppenheimer, renewed © 1981 by Robert B. Meyner; reprinted by permission of SIMON & SCHUSTER, Inc.

Prospects in the Arts and Sciences
From "The Open Mind", © 1955 by J. Robert Oppenheimer;
reprinted by permission of SIMON & SCHUSTER, Inc.

An Inward Look
Reprinted by permission of *Foreign Affairs*, January 1958; © 1958
by the Council on Foreign Relations, Inc.

Tradition and Discovery
Speech given at the University of Puerto Rico, 1960.

Progress in Freedom
Speech given at The Tenth Anniversary Conference Congress for
Cultural Freedom, Berlin, June 1960.

On Science and Culture
Reprinted by permission of *Encounter* (London).

The Power to Act . . .
Speech given at the Symposium on "Challenges to Democracy",
sponsored by the Center for the Study of Democratic Institutions
of the Fund for the Republic, Chicago, June 1963, under the title
"The Scientific Revolution and Its Effects on Democratic
Institutions."

A World without War
Speech given at the National Academy of Sciences Centennial,
Washington D.C., October 1963, under the title "Communication
and Comprehension of Scientific Knowledge."

L'Intime et le Commun . . .
Speech given at Rencontres Internationales de Genève, September 1964, under the title "L'Intime et le Commun"

To Live with Ourselves
Reprinted by permission of the U.S. Army Research Office,
United States Military Academy at West Point, N.Y.; from
"Principal Addresses of the 1965 U.S. Army National Junior
Science and Humanities Symposium".

Physics and Man's Understanding . . .
Speech given at the Smithsonian Institution Bicentennial Cele-
bration, Washington D.C., September 1965.

A Time in Need
Speech given at Princeton University, November 1966, under the
title "The Forbearance of Nations."